Universitext

Universitext

Universitext is a series of textbooks that presents material from a wide variety of mathematical disciplines at master's level and beyond. The books, often well class-tested by their author, may have an informal, personal even experimental approach to their subject matter. Some of the most successful and established books in the series have evolved through several editions, always following the evolution of teaching curricula, to very polished texts.

Thus as research topics trickle down into graduate-level teaching, first textbooks written for new, cutting-edge courses may make their way into *Universitext*.

For further volumes:
http://www.springer.com/series/223

Mak Trifković

Algebraic Theory of Quadratic Numbers

 Springer

Mak Trifković
Department of Math and Statistics
University of Victoria
Victoria, BC, Canada

ISSN 0172-5939   ISSN 2191-6675 (electronic)
ISBN 978-1-4614-7716-7   ISBN 978-1-4614-7717-4 (eBook)
DOI 10.1007/978-1-4614-7717-4
Springer New York Heidelberg Dordrecht London

Library of Congress Control Number: 2013941873

Mathematics Subject Classification: 11-01

Printed on acid-free paper

Springer is part of Springer Science+Business Media (www.springer.com)

Preface

Elementary arithmetic studies divisibility and factorization of ordinary integers. Algebraic number theory considers the same questions for algebraic numbers, solutions to polynomial equations with integer coefficients. In this setting, (un)fortunately, the uniqueness of prime factorization no longer holds. Remedying and measuring its failure is the starting point of algebraic number theory.

There are many excellent texts both on elementary and on algebraic number theory. However, I needed an intermediate-level book for an undergraduate course with the aim of imparting the flavor and beauty of algebraic number theory with minimal algebraic prerequisites. The notes for that course grew into this book. Restricting to quadratic numbers was a natural choice: hands-on examples abound, while the Galois theory is trivial to the point of invisibility. Indeed, concreteness and computation underpin the approach of the book. Computers are a great research tool; for learning, however, there is no substitute for getting one's hands dirty with calculations. A parallel emphasis is on the interaction between different branches of mathematics, all too often introduced to undergraduates as separate worlds. In this book, the student can see them joining forces to prove beautiful theorems. Linear and abstract algebra together rescue unique factorization; basic plane geometry is essential for proving that unique factorization never fails by much.

The prerequisites for this book are a knowledge of elementary number theory and a passing familiarity with ring theory. Its goal is to give the undergraduate a taste of algebraic number theory right after (or during) the first course in abstract algebra, and before learning Galois theory. Indeed, the book can serve as a rich source of examples for a ring theory course. A bright and brave freshman can read it, with some effort, even before the first course on rings: the necessary theory is covered, albeit briskly, in Chap. 2. Finally, the book can be a concrete supplement to a beginning graduate course in algebraic number theory.

Here's a brief outline of the contents.

Chapter 1 generalizes, to the extent possible, the arithmetic of $\mathbb{Z}$ to several specific quadratic fields. The examples are chosen to present the range of new phenomena in algebraic number theory. The definitions of a ring and a field are the only abstract algebra required here.

Chapter 2 is a self-contained but quick review of ring theory, with examples geared toward number theory.

Chapter 3 amplifies the standard theory of row- and column-reduction by restricting to matrices with entries in $\mathbb{Z}$.

Chapter 4 is the heart of the book. It develops algebraic number theory in a general quadratic field, and culminates in the proof of unique factorization of ideals.

Chapter 5 proves the finiteness of the ideal class group, with numerous computational examples.

Chapter 6 completes the picture of the arithmetic of quadratic fields by describing the group of units of a real quadratic field using the theory of continued fractions. This chapter and the next are arguably the most interesting ones, since their techniques and results don't generalize to higher-degree fields.

Chapter 7 goes back to the roots of algebraic number theory—binary quadratic forms. Forgoing the usual elementary presentation, we emphasize the connections with ideals. The main result of the chapter is a precise description of the identification of strict equivalence classes of binary quadratic forms and narrow ideal classes. This chapter has no extra prerequisites, but does require more mathematical maturity, as we introduce the language of group actions and commutative diagrams. Concreteness is not sacrificed, though: we cover in detail the algorithms for reducing both definite and indefinite forms. The section concludes with a presentation of Bhargava cubes, a recent development. The appendix assembles the results on the orders in quadratic fields, proved in the exercises throughout the book.

The terms being defined, either in formal definitions or in the running text, are **highlighted** for ease of finding. All propositions, definitions, displayed equations etc., are numbered in the same sequence. Starred exercises come with a hint at the back of the book.

Thanks go to my department, which supported me in giving small advanced classes; to the students therein, who helped me test the approach of the book; to my summer students, Chris Whitman for typing up these notes, and Jasper Wiart for carefully reading them and contributing to Chap. 7; and finally to Springer for their help and patience.

Victoria, BC, Canada   Mak Trifković

Notation

Continued Fractions

p_i/q_i	ith convergent of a continued fraction	p. 111
η_i	ith tail of the continued fraction of η	p. 109
m_i, v_i	Coefficients of $\eta_i = (m_i + \eta_{\mathcal{O}})/v_i$	p. 118

Quadratic Forms

$\mathcal{Q}, \mathcal{Q}_D, \mathcal{Q}_F$	Sets of quadratic forms	p. 140,140
$\mathcal{H}, \mathcal{H}_D, \mathcal{H}_F$	Sets of quadratic numbers	p. 140,140
$\mathcal{I}_F$	Set of oriented fractional ideals in F	p. 149
S, T	$\left[\begin{smallmatrix} 0 & -1 \\ 1 & 0 \end{smallmatrix}\right], \left[\begin{smallmatrix} 1 & 1 \\ 0 & 1 \end{smallmatrix}\right]$ as generators of $SL_2(\mathbb{Z})$	p. 145

Contents

List of Figures

Chapter 1
Examples

1.1 Review of Elementary Number Theory

When can we express a prime number as a sum of two squares? Let's start
by sorting the first dozen primes into those with such an expression, and the
rest:

$$p = a^2 + b^2 : \quad 2, 5, 13, 17, 29, 37 \ldots$$
$$p \neq a^2 + b^2 : \quad 3, 7, 11, 19, 23, 31 \ldots$$

Do you see a pattern?

This question was posed by Pierre de Fermat, a French eighteenth-century
mathematician. It may strike you as an unmotivated riddle, like his more
famous Last Theorem.[1] One of the joys of number theory is that even such
riddles often lead to beautiful, intrinsically interesting discoveries. Fermat's
question is our first example of the need to study divisibility, primes, and
factorizations in rings bigger than $\mathbb{Z}$.

In Ch. 1 we will look at concrete examples of such "higher" arithmetic.
To understand them you only need to know the definition of a ring and a field.
It helps to have some knowledge of ideals, but we leave an in-depth study of
those for the next chapter. By convention, all our rings are commutative and
have a multiplicative identity.

Recall that an integer $p \neq 0, \pm 1$ is **prime** when it has no integer divisors
other than ± 1 and $\pm p$. If a prime p is a sum of two squares, the equalities

$$p = a^2 + b^2 = (a + bi)(a - bi), \text{ with } a, b \in \mathbb{Z},$$

show that p acquires a nontrivial factorization once we allow factors from the
set

$$\mathbb{Z}[i] = \{a + bi : a, b \in \mathbb{Z}\}.$$

---

[1] Fermat's Last Theorem was for 350 years in fact just a conjecture: the equation $x^n + y^n = z^n$ has no integer solutions when $n \geq 3$. The proof of this conjecture, found only in the mid-1990's, is one of the great achievements of twentieth-century mathematics.

M. Trifković, *Algebraic Theory of Quadratic Numbers*, Universitext,
DOI 10.1007/978-1-4614-7717-4_1, © Springer Science+Business Media New York 2013

Like $\mathbb{Z}$, this is a ring: a set closed under addition, subtraction, and multiplication, but not necessarily division. It is in this new ring that we need to answer the basic arithmetical question, "how does p factor?" The goal of the book is to rigorously pose and answer such questions in quadratic rings, like $\mathbb{Z}[i]$. Those rings result from enlarging $\mathbb{Z}$ by a solution to a quadratic equation (in our example, $x^2 + 1 = 0$). General algebraic number theory undertakes the same task for polynomial equations of any degree. Most of its features, however, are already present in the quadratic case.

The main result of elementary arithmetic is the following well-known theorem.

1.1.1 Theorem (Unique Factorization in $\mathbb{Z}$). *Any integer other than 0 and ± 1 can be written as a product of primes, uniquely up to permuting the prime factors and changing their signs.*

The proof of Thm. 1.1.1 will be our template for extending arithmetic to rings such as $\mathbb{Z}[i]$. It proceeds through a chain of propositions.

1.1.2 Proposition (Division Algorithm). *Given $a, b \in \mathbb{Z}, b \neq 0$, there exist unique $q, r \in \mathbb{Z}$ such that*

$$a = qb + r \text{ and } 0 \leq r < |b| .$$

Proof. We prove the existence of q and r, leaving uniqueness as an exercise. Let's first assume that $b > 0$. Check that the set

$$\mathcal{S} = \{n \in \mathbb{Z} : n \geq 0 \text{ and } n = a - sb \text{ for some } s \in \mathbb{Z}\}$$

is nonempty. By the Well-Ordering Principle, $\mathcal{S}$ has a minimal element r, which must be of the form $r = a - qb$ for some $q \in \mathbb{Z}$. If we had $b \leq r$, then $0 \leq r - b = a - qb - b = a - (q+1)b$, so that $r - b \in \mathcal{S}$. We also have $r - b < r$ (since $b > 0$), which contradicts the choice of r as the minimum of $\mathcal{S}$.

If $b < 0$, we apply the preceding reasoning to a and $-b$ to find $q', r' \in \mathbb{Z}$ with $a = q'(-b) + r'$. We then put $q = -q', r = r'$.   ∎

Let $a, b \in \mathbb{Z}$, not both zero. The term **"greatest common divisor"** **(g.c.d.)** of a and b, denoted $\gcd(a, b)$, is self-explanatory. The next link in our chain of reasoning describes the important properties of the g.c.d.

1.1.3 Proposition (Euclid's Algorithm). *Let a and b be in $\mathbb{Z}$, not both zero. The greatest common divisor of a and b satisfies the following condition:*

(1.1.4)   *For any common divisor c of a and b, we have $c \mid \gcd(a, b)$.*

Moreover, there exist $r, s \in \mathbb{Z}$ such that $\gcd(a, b) = ra + sb$.

Proof. We prove the Proposition in a roundabout way that foreshadows the techniques we will use later. Since we want to express $\gcd(a, b)$ as a linear combination of a and b with integer coefficients, we consider the set of all such combinations,

$$I = \{ma + nb : m, n \in \mathbb{Z}\}.$$

Since a or b is non-zero, I has the smallest positive elements, $d = ra + sb$. We first show that d is a common divisor of a and b. Divide a by d with remainder: $a = qd + r$ with $0 \le r < d$. Then $r = a - qd = a - q(ra + sb) = (1 - qr)a + (-qs)b$ is an element of I. If $r > 0$, then r is a positive element of I smaller than d, contradicting our choice of d. Thus, $r = 0$ and $d \mid a$. An analogous argument shows that $d \mid b$.

Next, we show that d satisfies condition (1.1.4). Take $c \in \mathbb{Z}$ with $c \mid a$ and $c \mid b$. Then $c \mid ra + sb = d$. In particular, c is no bigger than d, so that $d = \gcd(a, b)$.   ∎

Check that the set I in the proof is closed under addition, and that it "absorbs multiplication": if $n \in \mathbb{Z}$ and $x \in I$, then $nx \in I$ also. In the terminology of Def. 2.2.1, I is our first example of an ideal (of the ring $\mathbb{Z}$).

1.1.5 Example. We'll illustrate the algorithm alluded to in the title of Prop. 1.1.3 (but not given in its proof!) by finding the g.c.d. of 598 and 273. The algorithm iterates division with a remainder, as follows:

$$598 = 2 \cdot 273 + 52$$
$$273 = 5 \cdot 52 + 13$$
$$52 = 4 \cdot 13$$

The divisor and the remainder in each line become the dividend and the divisor, respectively, in the following line. We get the g.c.d. as the last non-zero remainder, in this case $\gcd(598, 273) = 13$. Starting with the next-to-last line and going backwards, we calculate r and s promised by Prop. 1.1.3 by repeatedly expressing the remainder from the previous line:

$$13 = 273 - 5 \cdot 52 = 273 - 5 \cdot (598 - 2 \cdot 273) = 11 \cdot 273 - 5 \cdot 598. \qquad \square$$

Writing $\gcd(a, b)$ as a linear combination of a and b gives us a good algebraic handle on questions of divisibility. The proof of the next proposition is an example.

1.1.6 Proposition (Euclid's Lemma). *For any prime p and any $a, b \in \mathbb{Z}$, $p \mid ab$ implies $p \mid a$ or $p \mid b$.*

Proof. Assume that $p \mid ab$ and $p \nmid a$. Since p is prime, $\gcd(a, p)$ is 1 or p. In the latter case, we'd have $p = \gcd(a, p) \mid a$, contradicting the second assumption. Thus, $\gcd(a, p) = 1$, and Prop. 1.1.3 provides $r, s \in \mathbb{Z}$ such that $1 = rp + sa$. Then $b = b \cdot 1 = p(br) + (ab)s$ is divisible by p, by the assumption that $p \mid ab$.   ∎

We are now ready for the final link in the proof of Thm. 1.1.1. A simple inductive argument will prove the existence of a prime factorization. The uniqueness lies deeper, requiring Euclid's Lemma.

Proof of Unique Factorization in $\mathbb{Z}$. It is enough to prove that any integer greater than 1 is a product of positive primes, and that any two such factorizations differ only in the order of prime factors.

Existence of a prime divisor: Let n be an integer greater than 1, and let p be the smallest divisor of n which is greater than 1. We claim that p is prime. If not, it's divisible by some x with $1 < x < p$. By construction of p, those inequalities imply $x \nmid n$, contradicting $x \mid p \mid n$.

Existence of a prime factorization: We want to show that the set

$$\mathcal{S} = \{x \in \mathbb{N} \setminus \{1\} : x \text{ can't be written as a product of primes}\}$$

is empty. Otherwise, the Well-Ordering Principle guarantees a minimal element $n \in \mathcal{S}$. By the preceding paragraph, we can write $m = pn$ for some prime p. If $n = 1$, then $m = p$ would be a prime factorization, contradicting $m \in \mathcal{S}$. Thus, $1 < n < m$, and by definition of m we have $n \notin \mathcal{S}$. In other words, n does have a prime factorization $n = p_1 \cdots p_r$. But then $m = pp_1 \cdots p_r$ is a prime factorization of m, again contradicting $m \in \mathcal{S}$.

Uniqueness: Assume that we have two prime factorizations $n = p_1 \cdots p_r = q_1 \cdots q_s$. As p_1 is prime, Euclid's Lemma implies, after possibly permuting the factors, that $p_1 \mid q_1$. As q_1 is also prime, we must have $p_1 = q_1$. Cancelling gives two shorter factorizations $p_2 \cdots p_r = q_2 \cdots q_s$. We repeat the procedure until each prime in the first factorization is matched with one in the second. ∎

We would like to have an analog of unique factorization for **quadratic numbers**, solutions to equations of the form $ax^2 + bx + c = 0$ with $a, b, c \in \mathbb{Z}$. We will often need to solve such an equation modulo n, which we could do by plugging in for x all the elements in $\mathbb{Z}/n\mathbb{Z}$. A more efficient approach is given by the theory of quadratic residues. We recall it here without proof.

1.1.7 Definition. *Let $p \in \mathbb{N}$ be a positive odd prime, and let $a \in \mathbb{Z}$. We say that "a is a square mod p" when $a \equiv b^2 \pmod{p}$ for some $b \in \mathbb{Z}$. We define the Legendre symbol by*

$$\left(\frac{a}{p}\right) = \begin{cases} 1 & \text{if } a \text{ is a nonzero square mod } p \\ -1 & \text{if } a \text{ is not a square mod } p \\ 0 & \text{if } p \mid a \end{cases}$$

1.1.8 Example. Let's find "by hand" the Legendre symbols $\left(\frac{3}{11}\right), \left(\frac{7}{11}\right)$, and $\left(\frac{22}{11}\right)$. We compute all squares in $\mathbb{Z}/11\mathbb{Z}$:

x mod 11	0	±1	±2	±3	±4	±5
x^2 mod 11	0	1	4	9	5	3

From this we see that 3 is a square modulo 11, but 7 isn't: $\left(\frac{3}{11}\right) = 1$ and $\left(\frac{7}{11}\right) = -1$. By Def. 1.1.7, we have $\left(\frac{22}{11}\right) = 0$, as $11 \mid 22$. $\qquad\square$

The following theorem is an efficient alternative to calculating the Legendre symbol by brute force.

1.1.9 Theorem. *Let $a, b \in \mathbb{Z}$, and take $p, q \in \mathbb{N}$ to be distinct positive odd primes. The Legendre symbol satisfies the following properties:*

(a) $\left(\frac{a}{p}\right) \equiv a^{\frac{p-1}{2}} \pmod{p}$

(b) $\left(\frac{ab}{p}\right) = \left(\frac{a}{p}\right)\left(\frac{b}{p}\right)$

(c) *If $a \equiv b \mod p$, then $\left(\frac{a}{p}\right) = \left(\frac{b}{p}\right)$.*

(d) $\left(\frac{p}{q}\right) = \begin{cases} -\left(\frac{q}{p}\right) & \text{if } p \equiv q \equiv 3 \pmod{4} \\ \left(\frac{q}{p}\right) & \text{otherwise} \end{cases}$

(e) $\left(\frac{-1}{p}\right) = \begin{cases} 1 & \text{if } p \equiv 1 \pmod{4} \\ -1 & \text{if } p \equiv 3 \pmod{4} \end{cases}$

(f) $\left(\frac{2}{p}\right) = \begin{cases} 1 & \text{if } p \equiv \pm 1 \pmod{8} \\ -1 & \text{if } p \equiv \pm 3 \pmod{8} \end{cases}$

Part (d) of the theorem is usually referred to as the Law of Quadratic Reciprocity. Remember that the denominator in the Legendre symbol must be a positive prime. For a generalization that removes that condition, see Exer. 1.1.13.

1.1.10 Example. The following computation illustrates how to use the six properties of Thm. 1.1.9 to compute the Legendre symbol. With variations, we reduce the numerator modulo the denominator, then factor, apply reciprocity, and repeat. Each equality is justified by one of the properties (a)–(f) above, as indicated by the letter(s) above the equal signs.

$$\left(\frac{-249}{107}\right) \overset{b}{=} \left(\frac{-1}{107}\right)\left(\frac{249}{107}\right) \overset{e,c}{=} -\left(\frac{35}{107}\right) \overset{b}{=} -\left(\frac{5}{107}\right)\left(\frac{7}{107}\right)$$

$$\overset{d}{=} -\left(\frac{107}{5}\right)\left[-\left(\frac{107}{7}\right)\right] \overset{c}{=} \left(\frac{2}{5}\right)\left(\frac{2}{7}\right) \overset{f}{=} (-1) \cdot 1 = -1.$$

There are other ways of computing $\left(\frac{-249}{107}\right)$: for example, we could start by factoring -249. $\qquad\square$

We are now ready to start studying the arithmetic of quadratic numbers. Such a number, being a solution to $ax^2 + bx + c = 0$ for some $a, b, c \in \mathbb{Z}$, is given by

$$x = \frac{-b \pm \sqrt{D}}{2a}, \text{ where } D = b^2 - 4ac.$$

The quadratic formula suggests that we should work in a field containing $\sqrt{D}$. For efficiency, we may as well take the smallest one,

$$F = \mathbb{Q}[\sqrt{D}] = \{a + b\sqrt{D} : a, b \in \mathbb{Q}\},$$

pronounced "$\mathbb{Q}$-adjoin-$\sqrt{D}$." Any such field is termed a **quadratic field**.

We can always divide in a field. To ask interesting questions about divisibility, we should work in a natural subring $\mathcal{O} \subset F$ analogous $\mathbb{Z} \subset \mathbb{Q}$. Ideally, the elements of $\mathcal{O}$ should have unique factorization into "prime" elements, whatever that means. This is too much to ask for, as we will see in Sec. 1.5. There are two ways of dealing with the failure of unique factorization, corresponding to two of the main results of this book:

(a) While quadratic *numbers* don't necessarily have unique factorization, we will prove in Sec. 4.8 that each *ideal* of $\mathcal{O}$ can be written, essentially uniquely, as a product of prime ideals (all terms to be defined in due course). This is foreshadowed by the ambiguity of signs of prime factors in Thm. 1.1.1, since p_i and $-p_i$ generate the same ideal in $\mathbb{Z}$.
(b) We will define an abelian group called the ideal class group of F. It acts as a flag, being trivial precisely when $\mathcal{O}$ has unique factorization, and otherwise quantifying the extent of its failure. Thm. 5.1.2 shows that the ideal class group is finite, so that the failure of unique factorization is never too bad.

In the final chapter we will return to the late eighteenth-century roots of number theory. Our third main result, Thm. 7.8.1, gives an elementary interpretation of the ideal class group in terms of quadratic forms.

Exercises

1.1.1. Fill in the gaps in the proof of Prop. 1.1.2: prove that $\mathcal{S} \neq \emptyset$, and that q and r are unique.

1.1.2. Prove the following version of the division algorithm: given $a, b \in \mathbb{Z}$ with $b \neq 0$, there exist unique $q', r' \in \mathbb{Z}$ such that

$$a = q'b + r', \text{ and } -\left|\frac{b}{2}\right| < r' \le \left|\frac{b}{2}\right|.$$

1.1.3. Use Euclid's Algorithm to explicitly write $\gcd(a, b) = ra + sb$ for: (a) $a = 137, b = 715$, (b) $a = 10530, b = 3978$.

1.1.4. Let $n \in \mathbb{N}, n > 1$. Show that n is either prime, or it has a prime factor $\le \sqrt{n}$. Use this to find the prime factorizations of 5532 and 64476.

1.1.5. In any elementary number theory textbook, look up the following two beautiful arguments, known to the ancient Greeks:

(a) The Sieve of Eratosthenes; use it to write down all primes < 100.

(b) Euclid's argument for the existence of infinitely many primes.

1.1.6. Prove that the set $\mathbb{Q}[\sqrt{D}] = \{a + b\sqrt{D} : a, b \in \mathbb{Q}\}$ is a field with respect to the usual addition and multiplication of complex numbers.

1.1.7. Compute using Quadratic Reciprocity: (a) $\left(\frac{-270}{17}\right)$; (b) $\left(\frac{47}{53}\right)$; (c) $\left(\frac{35}{83}\right)$; (d) $\left(\frac{629}{821}\right)$; e) $\left(\frac{1245}{2347}\right)$.

1.1.8.* Let $p \in \mathbb{N}$ be an odd prime. Show that exactly half the nonzero elements of $\mathbb{Z}/p\mathbb{Z}$ are squares.

1.1.9.* Prove the Chinese Remainder Theorem: for any $n_1, n_2 \in \mathbb{N}$ with $\gcd(n_1, n_2) = 1$ and any $a_1, a_2 \in \mathbb{Z}$, there exists an $x_0 \in \mathbb{Z}$ simultaneously satisfying the congruences

$$x \equiv a_1 \pmod{n_1}, \quad x \equiv a_2 \pmod{n_2}.$$

Moreover, all such x are given by $x \equiv x_0 \pmod{n_1 n_2}$.

1.1.10.* In each of the three examples, find all $x \in \mathbb{Z}$ simultaneously satisfying all the congruences:

(a) $x \equiv 5 \pmod 9$, $\quad x \equiv 12 \pmod{17}$

(b) $x \equiv 1 \pmod 7$, $\quad x \equiv 2 \pmod 8$, $\quad x \equiv 3 \pmod 9$

(c) $x \equiv 7 \pmod{12}$, $\quad x \equiv 37 \pmod{45}$, $\quad x \equiv 27 \pmod{50}$

1.1.11. Let $n_1, n_2, \in \mathbb{N}$. Here is a version of the Chinese Remainder Theorem which doesn't require that $\gcd(n_1, n_2) = 1$. Take $a_1, a_2 \in \mathbb{Z}$ satisfying $a_1 \equiv a_2 \pmod{\gcd(n_1, n_2)}$. Show that there exists $x_0 \in \mathbb{Z}$ with $x_0 \equiv a_i \pmod{n_i}$ for $n = 1, 2$, and that this x_0 is unique modulo $\operatorname{lcm}(n_1, n_2)$.

1.1.12. We will prove that there are infinitely many primes in $\mathbb{N}$, using calculus.

(a) Let $s \in \mathbb{R}$. Use the integral comparison test to show that the sum $\sum_{n=1}^{\infty} \frac{1}{n^s}$ converges for $s > 1$, but diverges for $s = 1$. For $s > 1$, the sum of this series, denoted $\zeta(s)$, is the famous Riemann zeta function.

(b) Let $p > 1$. For which values of $s \in \mathbb{R}$, if any, does the following equality hold:

$$\frac{1}{1 - p^{-s}} = 1 + p^{-s} + p^{-2s} + p^{-3s} + \cdots?$$

(c) Let $T = \{p_1, \ldots, p_r\}$ be a finite set of primes. Use Unique Factorization to show that

$$\prod_{i=1}^{r} \left(1 - \frac{1}{p_i^s}\right)^{-1} = \sum_{n = p_1^{e_1} \ldots p_r^{e_r}} \frac{1}{n^s}, \quad \text{for } s > 0,$$

where the sum ranges over all $n \in \mathbb{N}$ which have no prime factors outside $p_1, \ldots, p_r$.

(d) Pass to the limit as $|T| \to \infty$ in c) to show that

$$\prod_{p \text{ prime}} \left(1 - \frac{1}{p^s}\right)^{-1} = \zeta(s), \text{ for } s > 1.$$

(e) Consider the limit $\lim_{s \to 1+} \zeta(s)$, and use parts (a) and (d) to conclude that there are infinitely many primes.

1.1.13. The definition of the Legendre symbol $\left(\frac{a}{p}\right)$ requires that the denominator be a positive odd prime. The Kronecker symbol, still denoted $\left(\frac{a}{n}\right)$, extends the Legendre symbol to remove these restrictions. Let $a, n \in \mathbb{Z}$ with $n \neq 0$, and let $n = up_1^{e_1} \cdots p_r^{e_r}$ be the prime factorization of n, with $u = \pm 1$ and $p_1, \ldots, p_r$ distinct positive primes. We then define

$$\left(\frac{a}{n}\right) = \left(\frac{a}{u}\right) \prod_{i=1}^{r} \left(\frac{a}{p_i}\right)^{e_i}.$$

The value of $\left(\frac{a}{p_i}\right)$ for p_i odd is the usual Legendre symbol. We put $\left(\frac{a}{1}\right) = 1$, and define

$$\left(\frac{a}{-1}\right) = \begin{cases} 1 & \text{if } a \geq 0 \\ -1 & \text{if } a < 0 \end{cases} \qquad \left(\frac{a}{2}\right) = \begin{cases} 0 & \text{if } 2 \mid a \\ 1 & \text{if } a \equiv \pm 1 \pmod 8 \\ -1 & \text{if } a \equiv \pm 3 \pmod 8 \end{cases}$$

(a) Prove that the Kronecker symbol is multiplicative in both variables: $\left(\frac{ab}{n}\right) = \left(\frac{a}{n}\right)\left(\frac{b}{n}\right)$ and $\left(\frac{a}{mn}\right) = \left(\frac{a}{m}\right)\left(\frac{a}{n}\right)$.

(b) If $m, n \in \mathbb{N}$, show that $\left(\frac{m}{n}\right) = \left(\frac{n}{m}\right)$ unless $\gcd(m, n) = 1$ and $m \equiv n \equiv 3$ (mod 4).

(c) Formulate and prove an analogous reciprocity law for arbitrary $m, n \in \mathbb{Z} \setminus 0$.

(d) If $m \equiv n \pmod a$, show that $\left(\frac{a}{m}\right) = \left(\frac{a}{n}\right)$.

1.2 The Field $\mathbb{Q}[i]$ and the Gauss Integers

Theorem 1.2.18 in this section answers Fermat's question, "when is a prime a sum of two squares?" As suggested in Sec. 1.1, the answer will come from studying arithmetic in the ring $\mathbb{Z}[i] = \{a + bi : a, b \in \mathbb{Z}\}$. In a sense that we'll make precise in Sec. 1.3, $\mathbb{Z}[i]$ is the best subring of the field $\mathbb{Q}[i] = \{a + bi : a, b \in \mathbb{Q}\}$ in which to do arithmetic, much like $\mathbb{Z}$ is for $\mathbb{Q}$. Indeed, the goal of this section is to prove that unique factorization holds in $\mathbb{Z}[i]$.

We will refer to elements of $\mathbb{Z}[i]$ as **Gauss integers**[2]. They form a two-dimensional, discrete, regular array of points in the complex plane, pictured in Fig. 1.1. It is an example of a **lattice**, a notion which we'll rigorously define and study in Ch. 3. The shaded square with vertices $0, 1, i$ and $1 + i$ is a **fundamental parallelogram** of this lattice: its translates by the points in the lattice tile the plane without overlap. Geometrically, Fermat's question asks: when is there a point in the lattice $\mathbb{Z}[i]$ on the circle of radius $\sqrt{p}$, centered at the origin? Exer. 5.2.1 will give us a purely geometric answer; for now, we pursue it algebraically.

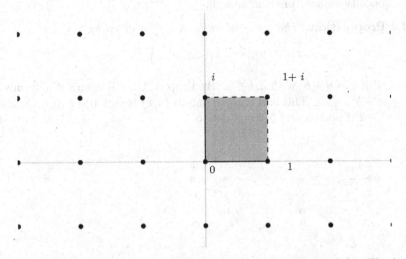

Fig. 1.1 The Gauss integers $\mathbb{Z}[i]$, with a fundamental parallelogram shaded in. The *dashed edges*, including the vertices $1, i$ and $1+i$, do not belong to the fundamental parallelogram. This ensures that any two of its translates by points in $\mathbb{Z}[i]$ are disjoint.

A **unit** of a ring R is an element with a multiplicative inverse. The units form a group under multiplication, denoted $R^{\times}$. For example, $F^{\times} = F \setminus 0$ for any field F. Moreover, $\mathbb{Z}^{\times} = \{\pm 1\}$, so that the prime factorization of an integer is unique up to multiplying the factors by units, and permuting them.

To find the units in $\mathbb{Z}[i]$, we use the following simple but versatile tool.

1.2.1 Definition. *The* **norm** *of* $\alpha = a + bi \in \mathbb{Q}[i]$ *is* $\mathrm{N}\alpha = \alpha\bar{\alpha} = a^2 + b^2 \in \mathbb{Q}$.

Note that $\mathrm{N}\alpha$ is always positive, being the square of the distance between α and the origin in the complex plane. Importantly, the norm is a homomorphism of multiplicative groups $\mathbb{Q}[i]^{\times} \to \mathbb{Q}^{\times}$. It sends $\mathbb{Z}[i]$ to $\mathbb{Z}$, thus reducing questions of divisibility in $\mathbb{Z}[i]$ to the corresponding, simpler ones in $\mathbb{Z}$.

---

[2] This name is slightly nonstandard; the literature usually refers to $\mathbb{Z}[i]$ as the ring of Gauss*ian* integers. I prefer to drop the *-ian* ending, as it is unnecessary, inconsistently applied, and awkward to pronounce with some non-English names.

1.2.2 Proposition. *For $\alpha, \beta \in \mathbb{Q}[i]$, $\mathrm{N}(\alpha\beta) = \mathrm{N}\alpha \cdot \mathrm{N}\beta$.*

Proof. $\mathrm{N}(\alpha\beta) = \alpha\beta \cdot \overline{\alpha\beta} = \alpha\bar{\alpha}\beta\bar{\beta} = \mathrm{N}\alpha \cdot \mathrm{N}\beta$.   ∎

1.2.3 Proposition. *An $\varepsilon \in \mathbb{Z}[i]$ is a unit if and only if $\mathrm{N}\varepsilon = \pm 1$.*

Proof. If $\varepsilon \in \mathbb{Z}[i]^{\times}$, there is a $v \in \mathbb{Z}[i]$ with $\varepsilon v = 1$. Taking the norm, we get $\mathrm{N}\varepsilon \cdot \mathrm{N}v = 1$, so that $\mathrm{N}\varepsilon \in \mathbb{Z}^{\times}$. Conversely, if $\mathrm{N}\varepsilon = \varepsilon\bar{\varepsilon} = \pm 1$, we find that $\varepsilon^{-1} = \pm\bar{\varepsilon} \in \mathbb{Z}[i]$, as desired.   ∎

All other quadratic fields have a norm homomorphism, so the preceding two propositions hold in them as well.

1.2.4 Proposition. *The group of units in $\mathbb{Z}[i]$ is given by*

$$\mathbb{Z}[i]^{\times} = \{1, -1, i, -i\}.$$

Proof. Put $\varepsilon = a + bi$ with $a, b \in \mathbb{Z}$. By Prop. 1.2.3, ε is a unit if and only if $\mathrm{N}\varepsilon = a^2 + b^2 = \pm 1$. This can happen only if (a, b) is $(\pm 1, 0)$ or $(0, \pm 1)$, since $a^2 + b^2 \geq 2$ if both a and b are non-zero.   ∎

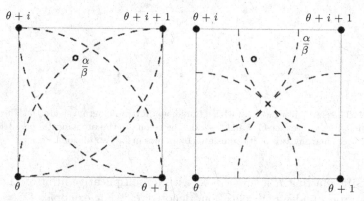

Fig. 1.2 α/β falls inside at least one of the circles of radius 1 centered at nearby lattice points (*left*). In fact, there is a point in $\mathbb{Z}[i]$ whose distance from α/β is at most $\sqrt{2}/2$ (*right*).

The division algorithm in $\mathbb{Z}$ allows us to divide with a remainder whose absolute value is small relative to that of the divisor. Replacing the absolute value by the norm, we get a division algorithm in $\mathbb{Z}[i]$.

1.2.5 Proposition. *Given $\alpha, \beta \in \mathbb{Z}[i], \beta \neq 0$, there exist $\kappa, \lambda \in \mathbb{Z}[i]$ such that $\alpha = \kappa\beta + \lambda$, with $\mathrm{N}\lambda < \mathrm{N}\beta$.*

Proof. We have already observed that the translates of the fundamental parallelogram by points in $\mathbb{Z}[i]$ tile the complex plane. The translate containing α/β is entirely covered by the four circles of radius 1 centered at its vertices (see Fig. 1.2). We take as our κ any vertex with a distance to α/β that is less than 1, i.e., $N(\alpha/\beta - \kappa) < 1$. Multiplying by $N\beta$, we get $N(\alpha - \kappa\beta) < N\beta$, so we put $\lambda = \alpha - \kappa\beta$. ∎

There could be several values of κ; the Proposition makes no uniqueness claims.

1.2.6 Example. Let's divide $7 + 3i$ by $2 + i$ with a remainder. In practice, we don't need a drawing like Fig. 1.2. Instead, we divide in $\mathbb{Q}[i]$:

$$\frac{7+3i}{2+i} = \frac{(7+3i)(2-i)}{5} = \frac{17-i}{5} = \frac{17}{5} - \frac{1}{5}i.$$

We round off the real and imaginary parts, and take $\kappa = 3+0i$. Consequently, $\lambda = \alpha - \kappa\beta = 7 + 3i - 3(2+i) = 1$. These choices satisfy the requirements of the division algorithm: $7 + 3i = 3(2+i) + 1$, and $N(1) = 1 < 5 = N(2+i)$. □

The definition of a prime number in $\mathbb{Z}$ has a direct generalization to $\mathbb{Z}[i]$. We change the term "prime" to "irreducible" to conform to general ring theory, where "prime" is reserved for a somewhat more restrictive notion (see Prop. 2.5.10).

1.2.7 Definition. *A nonunit element[3] $\pi \in \mathbb{Z}[i]$ is* **irreducible** *if $\pi = \alpha\beta$ implies $\alpha \in \mathbb{Z}[i]^\times$ or $\beta \in \mathbb{Z}[i]^\times$, for any $\alpha, \beta \in \mathbb{Z}[i]$.*

Our goal is to show that every nonunit $\alpha \in \mathbb{Z}[i]$ factors into irreducibles, and that any two such factorizations of α are equivalent, in the following sense.

1.2.8 Definition. *Let $\alpha = \pi_1\pi_2\cdots\pi_r = \pi_1'\pi_2'\cdots\pi_{r'}'$ be two factorizations of α into irreducible elements in $\mathbb{Z}[i]$. We say that the two factorizations are* **equivalent** *if they satisfy the following conditions:*

(a) $r = r'$, and
(b) There exists a permutation $\sigma\colon \{1,\ldots,r\} \to \{1,\ldots,r\}$, and units $\varepsilon_i \in \mathbb{Z}[i]^\times$, such that $\pi_i' = \varepsilon_i\pi_{\sigma(i)}$, for all $i \in \{1,\ldots,r\}$.

In plain English, two factorizations are equivalent if one can be obtained from the other by rearranging the factors and multiplying each by a unit.

The term "greatest common divisor" of two elements of $\mathbb{Z}[i]$ is no longer self-explanatory, since $\mathbb{Z}[i]$ does not have a natural total ordering. To define the g.c.d. in $\mathbb{Z}[i]$ and to prove it exists, we take our cue from Prop. 1.1.3.

1.2.9 Definition. *Let $\alpha, \beta \in \mathbb{Z}[i]$, not both zero. A* **greatest common divisor** *of α and β is any $\delta \in \mathbb{Z}[i]$ satisfying the following conditions:*

---

[3] Here π stands for a Gauss integer, *not* 3.141592....

(a) $\delta \mid \alpha$ and $\delta \mid \beta$, and
(b) For all $\gamma \in \mathbb{Z}[i]$, if $\gamma \mid \alpha$ and $\gamma \mid \beta$, then $\gamma \mid \delta$.

The g.c.d. is not uniquely defined. Indeed, if δ and δ' both satisfy the Definition, condition (b) shows that $\delta' = \varepsilon\delta$ for some unit ε.

1.2.10 Proposition. *Let $\alpha, \beta \in \mathbb{Z}[i]$, not both zero. There exists a g.c.d. δ of α and β. Any such δ can be written as $\delta = \varphi\alpha + \theta\beta$ for some $\varphi, \theta \in \mathbb{Z}[i]$.*

The last statement suggests looking for δ in the set

$$I = \{\mu\alpha + \nu\beta : \mu, \nu \in \mathbb{Z}[i]\}.$$

We will prove that I consists of all Gauss integers divisible by some $\delta \in I$. This requires only the following two properties of I:

- Additive closure: If $\xi, \zeta \in I$, then $\xi + \zeta \in I$; and
- Absorption: If $\tau \in \mathbb{Z}[i]$ and $\xi \in I$, then $\tau\xi \in I$ as well.

1.2.11 Lemma. *Let δ be an element of I with minimal nonzero norm. Then $I = \mathbb{Z}[i] \cdot \delta$, the set of all multiples of δ.*

Proof. Since $\delta \in I$, absorption implies that $\mathbb{Z}[i] \cdot \delta \subseteq I$. For the other inclusion we need to show that any $\eta \in I$ is divisible by δ. In other words: when we divide η by δ with remainder, we must show that the remainder is zero.

Take $\kappa, \lambda \in \mathbb{Z}[i]$ for which $\eta = \kappa\delta + \lambda$ and $N\lambda < N\delta$. Additive closure and absorption imply that $\lambda = \eta - \kappa\delta \in I$ also. If $\lambda \neq 0$, then λ is an element of I with nonzero norm smaller than δ, contradicting our choice of δ. We conclude that $\lambda = 0$, as desired. ∎

In the terminology of Ch. 2, additive closure and absorption make I an ideal of $\mathbb{Z}[i]$. Lemma 1.2.11 asserts that I is a principal ideal with generator δ.

Proof of Prop. 1.2.10. Let's check that the δ from Lemma 1.2.11 is a g.c.d. of α and β. As $\alpha, \beta \in I$, they're both multiples of δ. This is just condition (a) of Def. 1.2.9.

Since $\delta \in I$, we can write $\delta = \varphi\alpha + \theta\beta$ for some $\varphi, \theta \in \mathbb{Z}[i]$. Now take any common divisor γ of α and β. We have that $\gamma \mid \varphi\alpha + \theta\beta = \delta$, as required by condition (b). ∎

From this point on, the proof of Unique Factorization in $\mathbb{Z}[i]$ is entirely analogous to that in $\mathbb{Z}$. We just state the two remaining results and leave the details as an exercise.

1.2.12 Proposition (Euclid's Lemma for Gauss Integers). *For any irreducible element $\pi \in \mathbb{Z}[i]$ and any $\alpha, \beta \in \mathbb{Z}[i]$, $\pi \mid \alpha\beta$ implies $\pi \mid \alpha$ or $\pi \mid \beta$.*

1.2.13 Theorem (Unique Factorization for Gauss Integers). *Any nonunit element in $\mathbb{Z}[i]$ is a product of irreducible elements. Any two such factorizations are equivalent.*

1.2.14 Example. Theorem 1.2.13 asserts that the following two factorizations into irreducibles are equivalent:

$$1 + 3i = (1 + i)(2 + i) = (1 - 2i)(-1 + i).$$

To see this, we multiply each of the factors in the first factorization by a unit, $i(1 + i) = -1 + i, (-i)(2 + i) = 1 - 2i$, and then switch them. The two units multiply to $i(-i) = 1$, which was to be expected since both products are equal to $1 + 3i$.   □

The following simple sufficient condition shows that the factors in the Example are irreducible.

1.2.15 Proposition. *Let* $\alpha \in \mathbb{Z}[i]$. *If* $\mathrm{N}\alpha$ *is a prime in* $\mathbb{Z}$, *then* α *is irreducible in* $\mathbb{Z}[i]$.

Proof. Suppose $\alpha = \beta\gamma$. Taking the norm, we get $p = \mathrm{N}\alpha = \mathrm{N}\beta \cdot \mathrm{N}\gamma$. Since by assumption p is prime, we have, say, $\mathrm{N}\beta = \pm 1$. Then Prop. 1.2.3 implies that β is a unit in $\mathbb{Z}[i]$, so α is irreducible.   ∎

1.2.16 Example. We will see in Sec. 4.9 that the following examples represent all possible types of factorization of a prime in $\mathbb{Z}$ into irreducibles in $\mathbb{Z}[i]$.

(a) $2 = (-i)(1+i)^2$. By Prop. 1.2.15, $1+i$ is irreducible, as $\mathrm{N}(1+i) = 2$. The prime 2 thus becomes, up to a unit, the square of an irreducible in $\mathbb{Z}[i]$. This kind of behavior is rare; for Gauss integers, it happens only for 2.

(b) 3 remains irreducible in $\mathbb{Z}[i]$. Say $3 = \alpha\beta$ were a nontrivial factorization. Then $9 = \mathrm{N}(3) = \mathrm{N}\alpha \cdot \mathrm{N}\beta$, and since neither factor is a unit, we must have $\mathrm{N}\alpha = \mathrm{N}\beta = 3$. This can't happen: putting $\alpha = a + bi$, we must have $\mathrm{N}\alpha = a^2 + b^2 \equiv 0, 1$, or 2 (mod 4).

(c) $5 = (2 + i)(2 - i)$. The two factors are irreducible by Prop. 1.2.15. The integer prime 5 doesn't remain irreducible in $\mathbb{Z}[i]$, but "splits" into a product of two irreducible factors. Moreover, one irreducible factor is not a unit multiple of the other.   □

1.2.17 Example. Let's factor $\alpha = 165 + 111i$ into irreducibles in $\mathbb{Z}[i]$. If $\pi \in \mathbb{Z}[i]$ is an irreducible dividing α, then $\pi \mid \alpha\bar{\alpha} = \mathrm{N}\alpha = 39546 = 2 \cdot 3^2 \cdot 13^3$. By Euclid's Lemma, π must divide 2, 3 or 13. By unique factorization, π must be among their irreducible factors in $\mathbb{Z}[i]$:

$$2 = (-i)(1 + i)^2, \quad 3 \text{ is irreducible in } \mathbb{Z}[i], \quad 13 = (3 + 2i)(3 - 2i).$$

Thus, the factorization of α into irreducible elements is $\alpha = \varepsilon(1 + i)^a 3^b(3 + 2i)^c(3 - 2i)^d$ for some $a, b, c, d \geq 0$ and $\varepsilon \in \mathbb{Z}[i]^\times$. Consequently $\mathrm{N}\alpha = 2^a 9^b 13^{c+d}$, so that $a = 1, b = 1, c+d = 3$, and $\varepsilon(3+2i)^c(3-2i)^d = \alpha/(3+3i) = 46 - 9i$.

We could just test all four possible choices of c and d, but it's easier to be clever about it. If both c and d were at least 1, we'd have that $13 = (3+2i)(3-2i) \mid 46 - 9i$, which is clearly not the case. Thus, $46 - 9i$ is either $\varepsilon(3 + 2i)^3$ or $\varepsilon(3 - 2i)^3$. To decide which, it's enough to check whether $3 + 2i \mid 46 - 9i$ or $3 - 2i \mid 46 - 9i$. We find that $(46 - 9i)/(3 + 2i) = (120 + 119i)/13 \notin \mathbb{Z}[i]$, and $(46 - 9i)/(3 - 2i) = 12 + 5i \in \mathbb{Z}[i]$, so that finally

$$165 + 111i = i \cdot 3(1 + i)(3 - 2i)^3,$$

where we found $\varepsilon = i$ simply by dividing.   □

We now know enough about $\mathbb{Z}[i]$ to answer Fermat's question from Sec. 1.1.

1.2.18 Theorem. *A positive odd prime $p \in \mathbb{N}$ is a sum of two squares if and only if $p \equiv 1 \pmod 4$.*

Proof. Assume that there exist $a, b \in \mathbb{Z}$ with $p = a^2 + b^2$. As the only squares modulo 4 are 0 and 1, we must have $p \equiv 1 \pmod 4$.

Conversely, take a positive prime $p \equiv 1 \pmod 4$. By Quadratic Reciprocity, Thm. 1.1.9 (e), there exists an $n \in \mathbb{Z}$ such that $n^2 \equiv -1 \pmod p$, i.e. $p \mid n^2 + 1$. Factoring $n^2 + 1$ in $\mathbb{Z}[i]$, we get $p \mid (n + i)(n - i)$. Clearly, $p \nmid (n \pm i)$, otherwise the quotient Gauss integer would have imaginary part $\pm 1/p$. Thus, p, viewed as an element of $\mathbb{Z}[i]$, doesn't satisfy the conclusion of Euclid's Lemma and therefore can't be irreducible. Take a nontrivial factorization $p = \alpha\beta$ with $\mathrm{N}\alpha \neq 1 \neq \mathrm{N}\beta$. Apply the norm to get $p^2 = \mathrm{N}(p) = \mathrm{N}\alpha \cdot \mathrm{N}\beta$, which forces $\mathrm{N}\alpha = \mathrm{N}\beta = p$. Putting $\alpha = a + bi$, this translates to $p = a^2 + b^2$.   ∎

Theorem 1.2.18 is more than an answer to a curious question; it's essential to classifying the irreducibles in $\mathbb{Z}[i]$.

1.2.19 Theorem. *An element $\pi \in \mathbb{Z}[i]$ is irreducible if and only if one of the following conditions is met:*

(a) $\mathrm{N}\pi = p$ *is a prime in $\mathbb{N}$, necessarily $p \equiv 1 \pmod 4$ or $p = 2$; or*
(b) $\pi = \varepsilon p$, *for $\varepsilon \in \mathbb{Z}[i]^\times$ and $p \in \mathbb{N}$ is a positive prime congruent to 3 (mod 4).*

Proof. We first show that the Gauss integers listed in (a) and (b) are indeed irreducible. For (a), this follows from Prop. 1.2.15. For (b), assume that $p = \alpha\beta$, where $\alpha, \beta \notin \mathbb{Z}[i]^\times$. The proof of Thm. 1.2.18 shows that $\mathrm{N}\alpha = \mathrm{N}\beta = p$. But then we'd have $p \equiv 0, 1$ or 2 (mod 4), contradicting our assumption that $p \equiv 3 \pmod 4$.

Conversely, assume that $\pi \in \mathbb{Z}[i]$ is irreducible. Since $\pi \mid \pi\bar{\pi} = \mathrm{N}\pi$, Euclid's Lemma for $\mathbb{Z}[i]$ guarantees that π divides some integer prime factor p of $\mathrm{N}\pi \in \mathbb{Z}$. Then $\mathrm{N}\pi \mid \mathrm{N}p = p^2$, which happens if either:

(a) $N\pi = p$, so π is irreducible by Prop. 1.2.15; or,

(b) $N\pi = p^2$. Since π divides p, $\eta = p/\pi$ is in $\mathbb{Z}[i]$ and has norm

$$N\eta = \frac{Np}{N\pi} = \frac{p^2}{p^2} = 1.$$

This means that $\varepsilon = \eta^{-1}$ is a unit and $\pi = \varepsilon p$, as claimed. It remains to show that $p \equiv 3 \pmod 4$. If p were $\equiv 1 \pmod 4$, then by Thm. 1.2.18 we'd have $p = a^2 + b^2 = N\pi'$, for $\pi' = a + bi$ necessarily irreducible by Prop. 1.2.15. Now π' and π are distinct irreducible factors of p, since their norms are different. Hence π' divides a unit $\eta = p/\pi$, which is a contradiction. ∎

Exercises

1.2.1.* A triple $x, y, z \in \mathbb{N}$ is called Pythagorean if $x^2 + y^2 = z^2$ and $\gcd(x, y, z) = 1$. Show that all such triples are given by $x = u^2 - v^2, y = 2uv, z = u^2 + v^2$, where $u, v \in \mathbb{N}, u > v, \gcd(u, v) = 1$ and u, v not both odd.

1.2.2. Explain why the two following factorizations into irreducible elements don't contradict Unique Factorization in $\mathbb{Z}[i]$: $13 = (3 + 2i)(3 - 2i) = (2 + 3i)(2 - 3i)$?

1.2.3. Use the division algorithm in $\mathbb{Z}[i]$ to divide with remainder: (a) $3 + 4i$ by $2 - 3i$; (b) $17 - 17i$ by $7 + 6i$.

1.2.4. Use Euclid's Algorithm in $\mathbb{Z}[i]$ to find $\gcd(12 + i, 8 - i)$.

1.2.5. Factor the following into irreducibles in $\mathbb{Z}[i]$: (a) 23, (b) 41, (c) 3672, (d) $-6 + 18i$, (e) $145 - 58i$.

1.2.6. List, up to units, all irreducible elements of $\mathbb{Z}[i]$ with norm ≤ 50.

1.3 Quadratic Integers

The arithmetic in $\mathbb{Z}[i]$ is parallel to that in $\mathbb{Z}$ because both have a division algorithm. Its proof in Prop. 1.2.5 relied on the geometric fact that the Gauss integers are the vertices of a regular tiling of the complex plane by parallelograms. Such a subset of the complex plane is called a lattice.

1.3.1 Definition. *A lattice in $\mathbb{C}$ is a set of the form*

$$\mathbb{Z}\kappa + \mathbb{Z}\lambda = \{m\kappa + n\lambda : m, n \in \mathbb{Z}\},$$

where $\kappa, \lambda \in \mathbb{C} \setminus 0$ are noncollinear, i.e., $\lambda/\kappa \notin \mathbb{R}$.

We will define lattices somewhat more generally in Ch. 3. For now, you can think of a lattice as a discrete analog of a vector space with basis $\{\kappa, \lambda\}$.

What quadratic numbers other than those in $\mathbb{Z}[i]$ deserve to be considered a generalization of the usual integers, $\mathbb{Z}$? Inspired by the example of $\mathbb{Z}[i]$, let's look for subrings of $\mathbb{C}$ that are also lattices. Since a ring must contain 1, we're after lattices of the form $\mathbb{Z} + \mathbb{Z}\alpha$ which are closed under multiplication. This happens if and only if $\alpha^2 \in \mathbb{Z} + \mathbb{Z}\alpha$, giving us the following criterion:

(1.3.2) $\mathbb{Z} + \mathbb{Z}\alpha$ is a ring $\Leftrightarrow \alpha^2 + a\alpha + b = 0$ for some $a, b \in \mathbb{Z}$,

and suggesting our next definition.

1.3.3 Definition. *A* **quadratic integer** *is an $\alpha \in \mathbb{C}$ satisfying $\alpha^2 + a\alpha + b = 0$ for some $a, b \in \mathbb{Z}$.*

Two important remarks:

(a) The key requirement is that the polynomial in Def. 1.3.3 have coefficients in $\mathbb{Z}$ and be **monic**, i.e., have 1 as its leading coefficient. Otherwise, any polynomial with rational coefficients can be scaled up to one with coefficients in $\mathbb{Z}$ by clearing denominators.

(b) It is a nonobvious fact that the quadratic integers in the field $\mathbb{Q}[\sqrt{D}]$ form a ring. We will prove that by first explicitly describing them as a *set*, and then observing that this set is closed under addition, additive inverses, and multiplication.

The enlarging of $\mathbb{Z}$ by i to construct $\mathbb{Z}[i]$ generalizes to any ring R and any α an element of a larger ring, or a variable. The smallest ring containing both R and α is denoted by $R[\alpha]$, read "R-adjoin-α." Closure properties of a ring easily imply that

$$R[\alpha] = \{a_n\alpha^n + a_{n-1}\alpha^{n-1} + \cdots + a_0 : n \in \mathbb{Z}_{\geq 0}, a_k \in R \text{ for all } 0 \leq k \leq n\}.$$

If α is not a root of any polynomial with coefficients in R, monic or otherwise, then $R[\alpha]$ is the familiar **polynomial ring** in the variable α with coefficients in R.

If α satisfies a monic equation $\alpha^2 + a\alpha + b = 0$ with coefficients in R, we compute $\alpha^2 = -b - a\alpha$, $\alpha^3 = \alpha \cdot \alpha^2 = -b\alpha - a\alpha^2 = ab + (a^2 - b)\alpha$, etc. Since the leading coefficient in the equation satisfied by α is 1, every power of α, and therefore every element of $R[\alpha]$, is a linear combination of 1 and α with coefficients in R (an R-**linear** combination, for short). In particular, we have $\mathbb{Z}[\alpha] = \mathbb{Z} + \mathbb{Z}\alpha$, and we will use both notations.

Exercises

1.3.1. Let $p(x)$ be a polynomial with real coefficients, and $\alpha \in \mathbb{C}$. Show that $p(\alpha) = 0$ if and only if $p(\bar{\alpha}) = 0$. Deduce that if $p(\alpha) = 0$, then $p(x)$ is divisible by the quadratic polynomial $(x - \alpha)(x - \bar{\alpha})$, which has real coefficients.

1.3.2. Show that $(x-(a+b\sqrt{D})(x-(a-b\sqrt{D}))$ is the unique monic quadratic polynomial in $\mathbb{Q}[x]$ with the root $a + b\sqrt{D} \in \mathbb{Q}[\sqrt{D}]$.

1.3.3. Let $\alpha = a + bi, a, b \in \mathbb{Q}$. Use Exer. 1.3.2 to show that the set of all quadratic integers in $\mathbb{Q}[i]$ is precisely $\mathbb{Z}[i]$.

1.3.4. Let $p(x) = a_n x^n + a_{n-1} x^{n-1} + \cdots + a_0$ be a (not necessarily monic) polynomial with coefficients in $\mathbb{Z}$.

(a) If $p(r/s) = 0$ for $r, s \in \mathbb{Z}, \gcd(r, s) = 1$, show that $r \mid a_0$ and $s \mid a_n$.
(b) Show that $q \in \mathbb{Q}$ is a quadratic integer if and only if $q \in \mathbb{Z}$.

1.4 The Field $\mathbb{Q}[\sqrt{-3}]$ and the Eisenstein Integers

The notion of a quadratic integer gives us a framework for studying arithmetic in $\mathbb{Q}[\sqrt{-3}]$.

1.4.1 Proposition. *The set of all quadratic integers in $\mathbb{Q}[\sqrt{-3}]$ is $\mathbb{Z} + \mathbb{Z}\omega$, where $\omega = (1 + \sqrt{-3})/2$. In particular, this set is a ring, called the ring of* **Eisenstein integers**.

Proof. By Exer. 1.3.2, the unique monic quadratic polynomial with root $a + b\sqrt{-3}$ is

$$(x - (a + b\sqrt{-3}))(x - (a - b\sqrt{-3})) = x^2 - 2ax + (a^2 + 3b^2).$$

Let $\mathcal{O}$ be the set of all quadratic integers in $\mathbb{Q}[\sqrt{-3}]$. By Def. 1.3.3, $a + b\sqrt{-3} \in \mathcal{O}$ if and only if $2a = m, a^2 + 3b^2 = k$ for some $m, k \in \mathbb{Z}$. Write $b = r/s$ with $\gcd(r, s) = 1$. Then $12r^2/s^2 = 12b^2 = 4k - m^2 \in \mathbb{Z}$, which implies that $s^2 \mid 12$. That happens only for $s = 1$ or $s = 2$.

In either case, we can write $a = m/2$ and $b = n/2$ for some $m, n \in \mathbb{Z}$. Then $m^2/4 + 3n^2/4 = k \in \mathbb{Z}$, which implies $m^2 + 3n^2 \equiv m^2 - n^2 \equiv 0 \pmod{4}$. Since 0 and 1 are the only squares modulo 4, m and n must have the same parity: $m = n + 2l$ for some $l \in \mathbb{Z}$. Then

$$a + b\sqrt{-3} = \frac{m}{2} + \frac{n}{2}\sqrt{-3} = \frac{n + 2l}{2} + n\frac{\sqrt{-3}}{2} = l + n\frac{1 + \sqrt{-3}}{2} = l + n\omega,$$

showing that $\mathcal{O} \subseteq \mathbb{Z} + \mathbb{Z}\omega$. Check the reverse inclusion!

Since ω satisfies the monic polynomial $\omega^2 - \omega + 1 = 0$, the criterion (1.3.2) shows that the set $\mathbb{Z} + \mathbb{Z}\omega$ is a ring. ∎

The elements of $\mathbb{Z}[\omega] = \mathbb{Z} + \mathbb{Z}\omega$, depicted by bold dots in Fig. 1.3, form a lattice inside the complex plane.

1.4.2 Example. You might be wondering: why not simply work in $\mathbb{Z}[\sqrt{-3}] \subset \mathbb{Q}[\sqrt{-3}]$, by analogy with $\mathbb{Z}[i] \subset \mathbb{Q}[i]$? Observe that 4 has two inequivalent factorizations into irreducibles in $\in \mathbb{Z}[\sqrt{-3}]$:

$$4 = 2 \cdot 2 = (1 + \sqrt{-3})(1 - \sqrt{-3}).$$

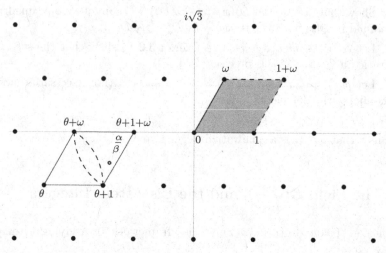

Fig. 1.3 The Eisenstein integers, with a fundamental parallelogram shaded in (*right*), and an illustration of the division algorithm (*left*).

While $(1 \pm \sqrt{-3})/2$ is not in $\mathbb{Z}[\sqrt{-3}]$, it is in fact a unit in $\mathbb{Z}[\omega]$. Viewed in this slightly bigger ring, the two apparently inequivalent factorizations of 4 differ only by units:

$$2 \cdot 2 = \left(2 \cdot \frac{1 + \sqrt{-3}}{2} \right) \cdot \left(\frac{1 - \sqrt{-3}}{2} \cdot 2 \right) = (1 + \sqrt{-3})(1 - \sqrt{-3}).$$

This is not an accident; we will soon prove that $\mathbb{Z}[\omega]$ does have unique factorization into irreducibles. In general, working with the biggest lattice–ring inside a quadratic field gives us the best chance of having some form of unique factorization, as we'll see in Sec. 4.8.   □

As with $\mathbb{Z}[i]$, we begin our study of $\mathbb{Z}[\omega]$ by finding its group of units. We again use the norm, defined for $\alpha = a + b\omega \in \mathbb{Q}[\sqrt{-3}]$ by

$$N\alpha = \alpha\bar{\alpha} = a^2 + ab + b^2.$$

As in Prop. 1.2.2, the norm is a homomorphism of multiplicative groups $\mathbb{Q}[\sqrt{-3}]^{\times} \to \mathbb{Q}^{\times}$.

1.4.3 Proposition. *The group of units of $\mathbb{Z}[\omega]$ is a cyclic group of order 6, generated by ω:*

$$\mathbb{Z}[\omega]^{\times} = \{1, \omega, \ldots, \omega^5\}.$$

Proof. The proof of Prop. 1.2.3 used only the existence of a multiplicative norm $\mathbb{Q}[i]^{\times} \to \mathbb{Q}^{\times}$, and is therefore valid for $\mathbb{Q}[\sqrt{-3}]$ as well. As in Prop. 1.2.4, we have $\varepsilon = a + b\omega \in \mathbb{Z}[\omega]^\times$ if and only if $\mathrm{N}\varepsilon = a^2 + ab + b^2 = 1$. Dividing the last equality by b^2, we get

$$\left(\frac{a}{b}\right)^2 + \frac{a}{b} + 1 = \frac{1}{b^2}.$$

It's easy to see that $x^2 + x + 1 \geq 3/4$ for all $x \in \mathbb{R}$. On the other hand, $1/b^2 \leq 1/4$ if $|b| > 1$, forcing $b = 0, \pm 1$. From this we solve for a to get the corresponding units in $\mathbb{Z}[\omega]$:

$$1 + 0 \cdot \omega = \omega^0, \qquad 0 + 1 \cdot \omega = \omega^1, \qquad -1 + 1 \cdot \omega = \omega^2$$
$$-1 + 0 \cdot \omega = \omega^3, \qquad 1 - 1 \cdot \omega = \omega^4, \qquad 0 - 1 \cdot \omega = \omega^5 \qquad \blacksquare$$

As promised, $\mathbb{Z}[\omega]$ has a division algorithm.

1.4.4 Proposition. *For any $\alpha, \beta \in \mathbb{Z}[\omega]$, $\beta \neq 0$, there exist $\kappa, \lambda \in \mathbb{Z}[\omega]$ such that $\alpha = \kappa\beta + \lambda$ and $\mathrm{N}\lambda < \mathrm{N}\beta$.*

Proof. We can find an element $\kappa \in \mathbb{Z}[\omega]$ such that $\mathrm{N}(\alpha/\beta - \kappa) < 1$. This is clear from Fig. 1.3: α/β lies in some translate of the fundamental parallelogram, and any such translate is be covered by two circles of radius 1. For the example in the figure, we may take $\kappa = \theta + \omega + 1$. Putting $\lambda = \alpha - \kappa\beta$, we have, as desired,

$$\mathrm{N}\lambda = \mathrm{N}(\alpha - \kappa\beta) = \mathrm{N}\beta \cdot \mathrm{N}\left(\frac{\alpha}{\beta} - \kappa\right) < \mathrm{N}\beta. \qquad \blacksquare$$

Now that we have a division algorithm, a chain of propositions entirely analogous to that in Sec. 1.2 proves the main result on the arithmetic of $\mathbb{Z}[\omega]$.

1.4.5 Theorem (Unique Factorization for Eisenstein Integers). *Any non-unit element of $\mathbb{Z}[\omega]$ can be written as a product of irreducible elements. Any two such factorizations are equivalent in the sense of Def. 1.2.8.*

Check the following classification of irreducible elements in $\mathbb{Z}[\omega]$:

1.4.6 Theorem. *An Eisenstein integer $\pi \in \mathbb{Z}[\omega]$ is irreducible if and only if one of the following conditions hold:*

(a) $\mathrm{N}\pi = p$ is a prime in $\mathbb{N}$, necessarily $p \equiv 1 \pmod 3$ or $p = 3$; or
(b) $\pi = \varepsilon p$, where $\varepsilon \in \mathbb{Z}[\omega]^\times$ and $p \in \mathbb{N}$ is prime with $p \equiv 2 \pmod 3$.

The proof is analogous to that of Thm. 1.2.19. The crucial step is the answer to a Fermat-type question, "which primes $p \in \mathbb{N}$ are of the form $p = x^2 + xy + y^2$, for some $x, y \in \mathbb{Z}$?"

Exercises

1.4.1. Show that $\mathbb{Z} + \mathbb{Z}\omega \subseteq \mathcal{O}$ by checking that for any $m, n \in \mathbb{Z}$, $m + n\omega$ is a root of a monic polynomial in $\mathbb{Z}[x]$.

1.4.2. Check the following identities, which enable you to easily compute in $\mathbb{Z}[\omega]$: (a) $\bar{\omega} = 1 - \omega$, (b) $\omega^2 = \omega - 1$; (c) $\omega^3 = -1$, (d) $N(a + b\omega) = a^2 + ab + b^2$.

1.4.3. We have the following factorizations in $\mathbb{Z}[\omega]$:

$$3 = -(\sqrt{-3})^2 = \frac{3 - \sqrt{-3}}{2} \cdot \frac{3 + \sqrt{-3}}{2}.$$

Check that the three factors involved are irreducible. How do you reconcile this with Unique Factorization in $\mathbb{Z}[\omega]$?

1.4.4.* Find the smallest constant $E \in \mathbb{R}$ for which the following version of the division algorithm holds: for any $\alpha, \beta \in \mathbb{Z}[\omega]$, $\beta \neq 0$, there exist $\gamma, \delta \in \mathbb{Z}[\omega]$ such that $\alpha = \beta\gamma + \delta$ and $N\delta \leq E \cdot N\beta$.

1.4.5. Use the division algorithm to divide with remainder: (a) $4 + 3\omega$ by $1 - 2\omega$, (b) $17 + 5\omega$ by $4 + 9\omega$.

1.4.6. Describe all irreducibles in $\mathbb{Z}[\omega]$ by adapting Thms. 1.2.18 and 1.2.19.

(a) Let $p \in \mathbb{N}$ be a positive odd prime. Prove that $\left(\frac{-3}{p}\right) = 1$ if and only if $p \equiv 1 \pmod{3}$.
(b) Find the necessary and sufficient conditions for a prime $p \in \mathbb{N}$ to be of the form $p = x^2 + xy + y^2$ for some $x, y \in \mathbb{Z}$.
(c) Prove the description of irreducibles in $\mathbb{Z}[\omega]$ given in Thm. 1.4.6.

1.4.7. List, up to units, all irreducible elements of $\mathbb{Z}[\omega]$ with norm ≤ 50.

1.4.8. Find the factorization into irreducibles in $\mathbb{Z}[\omega]$ of: (a) 3, (b) 5, (c) 143, (d) $9 - 12\omega$, (e) $-20 + 25\omega$.

1.5 The Field $\mathbb{Q}[\sqrt{-5}]$

The rings of integers in the fields $\mathbb{Q}$, $\mathbb{Q}[i]$, and $\mathbb{Q}[\sqrt{-3}]$ all have a division algorithm and therefore unique factorization. This optimistic pattern breaks down already for $\mathbb{Q}[\sqrt{-5}]$: we will see below that

(1.5.1) $6 = 2 \cdot 3 = (1 + \sqrt{-5})(1 - \sqrt{-5})$

are two inequivalent factorizations of 6 into irreducibles. Before we can rigorously verify that, we need to determine the ring of quadratic integers in $\mathbb{Q}[\sqrt{-5}]$, and its units.

1.5.2 Proposition. *The set of quadratic integers in $\mathbb{Q}[\sqrt{-5}]$ is $\mathbb{Z} + \mathbb{Z}\sqrt{-5}$, which is a ring.*

Proof. It's easy to check that all elements of $\mathbb{Z} + \mathbb{Z}\sqrt{-5}$ are quadratic integers. Conversely, assume $a + b\sqrt{-5} \in \mathbb{Q}[\sqrt{-5}]$ is a quadratic integer, so that the quadratic equation it satisfies, $x^2 - 2ax + (a^2 + 5b^2) = 0$, has coefficients in $\mathbb{Z}$. Then $a = m/2$ for some $m \in \mathbb{Z}$ and $a^2 + 5b^2 = (m^2 + 20b^2)/4$ is in $\mathbb{Z}$. This implies that $20b^2 = 20r^2/s^2$ is in $\mathbb{Z}$, where we wrote $b = r/s$ in lowest terms. As $\gcd(r,s) = 1$, this only happens if $s^2 \mid 20$. Thus, $s = 1$ or $s = 2$.

Either way, we can write $b = n/2$ for some $n \in \mathbb{Z}$, and get $a^2 + 5b^2 = (m^2 + 5n^2)/4 \in \mathbb{Z}$. Then $m^2 + n^2 \equiv m^2 + 5n^2 \equiv 0 \pmod{4}$. As the only squares modulo 4 are 0 and 1, the last congruence is possible only when both m and n are even. Consequently, $a = m/2$ and $b = n/2$ are both in $\mathbb{Z}$, so that the quadratic integer $a + b\sqrt{-5}$ is in $\mathbb{Z}[\sqrt{-5}]$.

Finally, $\mathbb{Z} + \mathbb{Z}\sqrt{-5} = \mathbb{Z}[\sqrt{-5}]$ is a ring by criterion (1.3.2), since $\sqrt{-5}$ satisfies the monic equation $x^2 + 5 = 0$. ∎

1.5.3 Proposition. $\mathbb{Z}[\sqrt{-5}]^\times = \{\pm 1\}$.

Proof. As in the proofs of Prop. 1.2.4 and Prop. 1.4.3, we look for $\varepsilon = a + b\sqrt{-5} \in \mathbb{Z}[\sqrt{-5}]$ such that $\mathrm{N}\varepsilon = a^2 + 5b^2 = \pm 1$. As $\mathrm{N}\varepsilon \geq 5$ when $b \neq 0$, we must have $b = 0$ and $\varepsilon = a = \pm 1$, as claimed. ∎

1.5.4 Example. We now show that (1.5.1) gives two inequivalent factorizations of 6 into irreducible elements in $\mathbb{Z}[\sqrt{-5}]$. Since the only units are ± 1, the factors in the first factorization aren't unit multiples of those in the second. It remains to check that $2, 3, (1 + \sqrt{-5})$ and $(1 - \sqrt{-5})$ are all irreducible in $\mathbb{Z}[\sqrt{-5}]$.

Suppose that $2 = \alpha\beta$, where $\alpha, \beta \in \mathbb{Z}[\sqrt{-5}]$ are both nonunits, i.e., $\mathrm{N}\alpha, \mathrm{N}\beta > 1$. Since $\mathrm{N}\alpha \cdot \mathrm{N}\beta = \mathrm{N}(2) = 4$, we must have $\mathrm{N}\alpha = \mathrm{N}\beta = 2$. Putting $\alpha = a + b\sqrt{-5}$, we get $a^2 + 5b^2 = 2$ and $a^2 \equiv 2 \pmod{5}$, which is impossible as the only squares modulo 5 are $0, 1$ and 4. Therefore, 2 is irreducible in $\mathbb{Z}[\sqrt{-5}]$. The proof for 3 is analogous.

If now $(1 \pm \sqrt{-5}) = \alpha\beta$ is a non-trivial factorization, we have $\mathrm{N}\alpha \cdot \mathrm{N}\beta = \mathrm{N}(1 \pm \sqrt{-5}) = 6$. Up to renaming, we must have $\mathrm{N}\alpha = 2$ and $\mathrm{N}\beta = 3$, both impossible by the preceding paragraph. □

Ernst Kummer was, in the mid-nineteenth century, the first to imagine a way of salvaging unique factorization by introducing "ideal numbers." We discuss them for motivation, so we won't define them formally. We simply treat them as objects for which statements about multiplication and divisibility make sense.

In our case, imagine there existed irreducible ideal numbers P_1, P_2, P_3 and P_4 such that

$$2 = P_1 P_2 \qquad\qquad 3 = P_3 P_4$$
$$1 + \sqrt{-5} = P_1 P_3 \qquad\qquad 1 - \sqrt{-5} = P_2 P_4.$$

Then the two factorizations of 6 in (1.5.1) would no longer have irreducible factors, hence wouldn't be required to be equivalent: after all, $24 = 6 \cdot 4 = 3 \cdot 8$ is not a counterexample to unique factorization in $\mathbb{Z}$. Rather, they would really just be rearrangements of a single factorization $6 = P_1 P_2 P_3 P_4$:

$$
\begin{array}{ccccccc}
6 = & 2 & \cdot & 3 & = (1 + \sqrt{-5}) \cdot (1 - \sqrt{-5}) \\
 & \| & & \| & \| & \| \\
(P_1 P_2) \cdot (P_3 P_4) = & (P_1 P_3) & \cdot & (P_2 P_4)
\end{array}
$$

So, how to define these ideal numbers? We need to make sense of multiplicative statements involving them, such as $A \mid \alpha$, where A is an ideal number and $\alpha \in \mathbb{Z}[\sqrt{-5}]$. Any divisibility worth the name should satisfy, for any $\alpha, \beta, \xi \in \mathbb{Z}[\sqrt{-5}]$,

$$A \mid \alpha, A \mid \beta \Rightarrow A \mid \alpha + \beta$$
$$A \mid \alpha \Rightarrow A \mid \xi \alpha.$$

In terms of the set $\mathcal{A} = \{\alpha \in \mathbb{Z}[\sqrt{-5}] : A \mid \alpha\}$, these conditions simply state that $\mathcal{A}$ is an ideal in the ring $\mathbb{Z}[\sqrt{-5}]$:

$$\alpha, \beta \in \mathcal{A} \Rightarrow \alpha + \beta \in \mathcal{A}$$
$$\xi \in \mathbb{Z}[\sqrt{-5}] \text{ and } \alpha \in \mathcal{A} \Rightarrow \xi \alpha \in \mathcal{A}.$$

Richard Dedekind, in the second half of the nineteenth century, made the leap from Kummer's heuristics of ideal numbers to rigorously defined ideals as the proper object for the study of factorizations. The culmination of his theory is the following theorem.

1.5.5 Theorem (Unique Factorization of Ideals). *Let $F = \mathbb{Q}[\sqrt{D}]$ be a quadratic field and $\mathcal{O}$ its ring of integers. For any ideal I of $\mathcal{O}$, other than 0 and $\mathcal{O}$, there exist prime ideals $P_1, P_2, \ldots P_n$ of $\mathcal{O}$, not necessarily distinct, such that $I = P_1 P_2 \ldots P_n$. This factorization is unique up to permutation of factors.*

In Ch. 2 we will develop the background material for the statement and proof of Thm. 1.5.5. In particular, we will rigorously define prime ideals and ideal multiplication.

Exercises

1.5.1.* By (1.5.1), 6 has two inequivalent factorizations into irreducible elements in $\mathbb{Z}[\sqrt{-5}]$. This ring therefore can't have a division algorithm. See what happens when you try to find $\gamma, \delta \in \mathbb{Z}[\sqrt{-5}]$ such that $3 = (1 - \sqrt{-5})\gamma + \delta$ with $N\delta < N(1 - \sqrt{-5})$.

1.5.2. Find all inequivalent factorizations of 21 into irreducible elements in $\mathbb{Z}[\sqrt{-5}]$.

1.5.3. Find two inequivalent factorizations of $44 + 10\sqrt{-5}$ into irreducible elements in $\mathbb{Z}[\sqrt{-5}]$.

1.5.4. Does unique factorization hold in $\mathbb{Z}[\sqrt{11}]$?

1.5.5. Study the field $F = \mathbb{Q}[\sqrt{-2}]$ along the lines of Secs. 1.2 and 1.4.

(a) Find the ring of integers $\mathcal{O}$ in $\mathbb{Q}[\sqrt{-2}]$.
(b) Show that a division algorithm holds in $\mathcal{O}$.
(c) Let $p \in \mathbb{N}$ be an odd prime. Use Quadratic Reciprocity (Thm. 1.1.9) to show that -2 is a square modulo p if and only if $p \equiv 1$ or 3 (mod 8).
(d) Use part (c) to describe all irreducible elements in $\mathcal{O}$, analogously to Thm. 1.2.19 and Exer. 1.4.6.

1.6 The Field $\mathbb{Q}[\sqrt{319}]$

The essential novelty of this example compared to the previous three is that the field $\mathbb{Q}[\sqrt{319}]$ is a **real quadratic field**: it's contained in $\mathbb{R}$, rather than merely in $\mathbb{C}$. By contrast, $\mathbb{Q}[i], \mathbb{Q}[\sqrt{-3}]$ and $\mathbb{Q}[\sqrt{-5}]$ are termed **imaginary quadratic fields**. In many ways, the two are similar. For instance, determining the ring of integers of $\mathbb{Q}[\sqrt{319}]$ is similar to Props. 1.4.1 and 1.5.2.

1.6.1 Proposition. *The set of quadratic integers in $\mathbb{Q}[\sqrt{319}]$ is $\mathbb{Z} + \mathbb{Z}\sqrt{319}$, which is a ring.*

Given an $\alpha = a + b\sqrt{319}$, we define its **conjugate** in $\mathbb{Q}[\sqrt{319}]$ by the formula $\bar{\alpha} = a - b\sqrt{319}$. Conjugation preserves addition and multiplication. The norm map, defined as before by $N\alpha = \alpha\bar{\alpha} = a^2 - 319b^2$, is a multiplicative homomorphism $\mathbb{Q}[\sqrt{319}]^\times \to \mathbb{Q}^\times$.

Real quadratic fields have more complicated structure than their imaginary friends. In $\mathbb{Z}[\sqrt{-5}]$, which is typical, the norm $N(a + b\sqrt{-5}) = a^2 + 5b^2$ is always positive and increases as we increase $|a|$ or $|b|$. The equation $a^2 + 5b^2 = n$ therefore has only finitely many solutions, and none when $n < 0$. Moreover, all solutions can be found by a finite search (e.g., by trying all a, b in the range $|a| \leq \sqrt{n}, |b| \leq \sqrt{n/5}$). By contrast, $N(a + b\sqrt{319}) = a^2 - 319b^2$ can be of either sign, since the terms a^2 and $-319b^2$ pull in opposite directions. This is at the root of all difficulties in studying arithmetic of real quadratic fields.

For example, to find all units along the lines of Props. 1.2.4, 1.4.3 and 1.5.3, we would have to find integer solutions to **Pell's equation**, $x^2 - 319y^2 = \pm 1$. There are infinitely many of them, the one with the smallest positive x being

$$x = 12901780, \quad y = 722361.$$

We're unlikely to find these numbers by trial and error. In fact, we chose to work in $\mathbb{Q}[\sqrt{319}]$ since their size illustrates the need for an algorithm which solves the general Pell's equation, $x^2 - Dy^2 = \pm 1$. Such an algorithm awaits in Ch. 6.

The unit $\varepsilon = 12901780 + 722361\sqrt{319}$, is called the **fundamental unit** of $\mathbb{Z}[\sqrt{319}]$, because we will show that

$$\mathbb{Z}[\sqrt{319}]^\times = \{\pm \varepsilon^n : n \in \mathbb{Z}\}.$$

The group of units of every real quadratic fields has this form. In particular, unlike in the complex case (see Prop. 4.2.7), real quadratic fields have infinitely many units.

The rings of integers in $\mathbb{Q}[i], \mathbb{Q}[\sqrt{-3}]$, and $\mathbb{Q}[\sqrt{-5}]$ are all lattices in $\mathbb{C}$. By contrast, in Exer. 1.6.4 you will show that $\mathbb{Z}[\sqrt{319}]$ is a dense subset of the real line: there is no hope of proving the division algorithm by tiling a plane with parallelograms. In Sec. 5.3 we will learn how to think of the ring of integers of a real quadratic field as a lattice in a plane. Even that won't help with $\mathbb{Z}[\sqrt{319}]$, since it in fact doesn't have unique factorization.

1.6.2 Example. We claim that the equalities

$$2747 = 47 \cdot 61 = -(2 + 3\sqrt{319})(2 - 3\sqrt{319})$$

give two inequivalent factorizations of 2747 into irreducible elements in $\mathbb{Z}[\sqrt{319}]$.

Suppose 47 is not irreducible, so that $47 = \alpha\beta$ for some $N\alpha, N\beta \neq \pm 1$. As in Ex. 1.5.4, this implies $N\alpha = N\beta = \pm 47$. Put $\alpha = a + b\sqrt{319}$ and consider $a^2 - 319b^2 = \pm 47$. Reducing modulo 29 (because $319 = 11 \cdot 29$), we get that ± 47 is a square modulo 29, which isn't true:

$$\left(\frac{\pm 47}{29}\right) = \left(\frac{47}{29}\right) = \left(\frac{-11}{29}\right) = \left(\frac{11}{29}\right) = \left(\frac{29}{11}\right) = \left(\frac{7}{11}\right) = -\left(\frac{11}{7}\right) = -\left(\frac{4}{7}\right) = -1.$$

The argument for 61 and $2 \pm 3\sqrt{319}$ is similar.

Finally, these two factorizations don't differ by units: the norms of the factors in the first factorization (47^2 and 61^2) are different from the norms of the factors in the second (both -2747). $\square$

Exercises

1.6.1. Prove that the equation $x^2 - 319y^2 = -1$ has no integer solutions.

1.6.2. Prove Prop. 1.6.1.

1.6.3. Verify the quadratic reciprocity computation in Ex. 1.6.2.

1.6.4. Denote by $(\eta) = \eta - \lfloor \eta \rfloor \in (0, 1)$ the fractional part of an irrational number $\eta \in \mathbb{R}$, and consider the set $\{(k\eta) : k \in \mathbb{Z}\} \subset (0, 1)$. By dividing the unit interval into n equal subintervals and applying the Pigeonhole Principle, find $a, b \in \mathbb{Z}$ with $|a + b\eta| < 1/n$. Conclude that $\mathbb{Z} + \mathbb{Z}\eta$; hence, in particular, $\mathbb{Z}[\sqrt{D}]$, is dense in $\mathbb{R}$.

1.6.5. Each of the following equations has a small integer solution. Find it by trial and error, or a computer search:

(a) $x^2 - 15y^2 = 1$
(b) $x^2 - 7y^2 = 1$
(c) $x^2 - 170y^2 = -1$
(d) $x^2 - 170y^2 = 1$

1.6.6. For the unit $\varepsilon = a_1 + b_1\sqrt{319}$, put $\varepsilon^n = a_n + b_n\sqrt{319}$. Give a recursive formula expressing a_{n+1}, b_{n+1} in terms of a_n, b_n, a_1, and b_1. This recurrence gives an elementary proof that the equation $x^2 - 319y^2 = 1$ has infinitely many solutions.

1.6.7.* Show that if the equation $x^2 - 319y^2 = n$ has one solution, it has infinitely many.

1.6.8.* Prove that there is no *continuous* function $c : \mathbb{R} \to \mathbb{R}$ that restricts to the conjugation on $\mathbb{Q}[\sqrt{319}]$, i.e., for which $c(a + b\sqrt{319}) = a - b\sqrt{319}$ for all $a, b \in \mathbb{Q}$.

Chapter 2
A Crash Course in Ring Theory

2.1 Basic Definitions

In $\mathbb{Z}$ we can add, subtract, and multiply without restrictions, but we can't always divide. That is what makes questions of divisibility and factorization interesting. To do arithmetic in more general number systems, we abstract these basic properties of $\mathbb{Z}$ to get the definition of a ring.

2.1.1 Definition. *A* **ring** $(R, +, \cdot)$ *is a set* R *with two binary operations (usually termed addition and multiplication) satisfying the following three axioms:*

(a) *$(R, +)$ is an abelian group with identity element 0_R.*
(b) *The operation $\cdot$ is associative, commutative and has an identity 1_R, such that $1_R \cdot a = a$, for all $a \in R$;*
(c) *Multiplication distributes over addition: for all $a, b, c \in R$, we have $a \cdot (b + c) = (a \cdot b) + (a \cdot c)$.*

A **subring** of R is a subset $S \subseteq R$ that contains 1_R and is closed under addition, additive inverses, and multiplication. This makes S a ring in its own right under the operations inherited from R, with $1_S = 1_R$. Unless confusion is possible, we'll drop the subscript and write 0 and 1 for 0_R and 1_R.

2.1.2 Definition. *A* **unit** *in R is an element $a \in R$ for which there exists a $b \in R$ such that $ab = 1$.*

The b in the Definition is unique and is called the multiplicative inverse of a, denoted a^{-1}. The set $R^\times$ of all units is a group under multiplication.

2.1.3 Example. Let $R = \mathbb{Z}[\sqrt{2}]$. It's easy to see that $\varepsilon = 1 + \sqrt{2}$ is a unit in R:

$$-N\varepsilon = \varepsilon(-\bar{\varepsilon}) = (1 + \sqrt{2})(-1 + \sqrt{2}) = 1.$$

We claim that the powers of ε are all distinct, and thus form an infinite set of units in $\mathbb{Z}[\sqrt{2}]$. Indeed, if $\varepsilon^k = \varepsilon^l$, then $\varepsilon^{k-l} = 1$. This can only be true

M. Trifković, *Algebraic Theory of Quadratic Numbers*, Universitext,
DOI 10.1007/978-1-4614-7717-4_2, © Springer Science+Business Media New York 2013

if $k = l$: otherwise, ε would be one of the two real roots of 1, namely 1 or -1. $\qquad\square$

Particularly simple and important are rings such as $\mathbb{Q}, \mathbb{R}$, and $\mathbb{C}$, in which we can divide by any nonzero element.

2.1.4 Definition. *A ring F is a* **field** *if every nonzero element is invertible, i.e., $F^\times = F \setminus 0$.*

The focus of this book will be fields of the form $\mathbb{Q}[\sqrt{D}]$.

Exercises

2.1.1. Decide which of the following are subrings of $\mathbb{C}$. Do you see a pattern?

(a) $\mathbb{Z} + \mathbb{Z}\frac{1+\sqrt{2}}{2}$

(b) $\mathbb{Z} + \mathbb{Z}\frac{1+\sqrt{3}}{2}$

(c) $\mathbb{Z} + \mathbb{Z}\frac{1+\sqrt{5}}{2}$

(d) $\mathbb{Z} + \mathbb{Z}\frac{1+\sqrt{-7}}{2}$

(e) $\mathbb{Z} + \mathbb{Z}\frac{1+\sqrt{-11}}{2}$

2.1.2. Let R and S be rings. Show that componentwise addition and multiplication make $R \times S = \{(r,s) : r \in R, s \in S\}$ into a ring.

2.1.3. Prove that the ring $\mathbb{Z}/n\mathbb{Z}$ is a field if and only if $n = p$, a prime. When we want to emphasize that we are considering $\mathbb{Z}/p\mathbb{Z}$ as a field, rather than merely an abelian group, we write $\mathbb{F}_p$ instead of $\mathbb{Z}/p\mathbb{Z}$.

2.1.4. Let R be a ring. For $n \in \mathbb{Z}$ and $a \in R$, we formally define the following intuitive operation: $n \cdot a = \underbrace{a + \cdots + a}_{n \text{ times}}$ if $n > 0$, $0 \cdot a = 0_R$, and $n \cdot a = (-n) \cdot (-a)$ if $n < 0$. We define the **characteristic of** R, denoted char R, as follows. If $n \cdot 1_R \neq 0_R$ for all $n \in \mathbb{Z} \setminus 0$, we put char $R = 0$. Otherwise, char R is the smallest positive integer n for which $n \cdot 1_R = 0$.

(a) Find the characteristics of the following rings: (a) $\mathbb{C}$; (b) $\mathbb{Z}/n\mathbb{Z}$; (c) $(\mathbb{Z}/n\mathbb{Z})[x]$; (d) $\mathbb{Z}/n\mathbb{Z} \times \mathbb{Z}/m\mathbb{Z}$.

(b) For a field F, prove that char F is 0 or a prime number.

2.1.5. Here we get to know a noncommutative ring that will make a cameo appearance in our investigations.

(a) For $k \in \mathbb{N}$, let $M_{k \times k}(\mathbb{Z})$ be the set of $k \times k$ matrices with entries in $\mathbb{Z}$, equipped with the usual matrix addition and multiplication. Show that $(M_{k \times k}(\mathbb{Z}), +, \cdot)$ satisfies all the ring axioms except the commutativity of multiplication.

(b) A unit in $M_{k \times k}(\mathbb{Z})$ is any a with a *two-sided* inverse: $ab = ba = 1$ for some $k \times k$ matrix b. Denote the unit group $M_{k \times k}(\mathbb{Z})^{\times}$ by $\mathrm{GL}_k(\mathbb{Z})$. Prove that

$$GL_2(\mathbb{Z}) = \left\{ \begin{bmatrix} a & b \\ c & d \end{bmatrix} : a, b, c, d \in \mathbb{Z}, \quad ad - bc = \pm 1 \right\}.$$

(c) Find a necessary and sufficient condition for $a, b \in \mathbb{Z}$ to appear in the top row of a matrix $\begin{bmatrix} a & b \\ c & d \end{bmatrix} \in GL_2(\mathbb{Z})$.

2.2 Ideals, Homomorphisms, and Quotients

In Ch. 1 we hinted at the importance of ideals to arithmetic. Their role in ring theory is analogous to that of normal subgroups in group theory.

2.2.1 Definition. *Let R be a ring. An **ideal** $I \subseteq R$ is an additive subgroup that absorbs multiplication: if $a \in R$ and $x \in I$, then $ax \in I$.*

Most of the ideals in this book will be of the form $\mathbb{Z}\alpha + \mathbb{Z}\beta = \{m\alpha + n\beta : m, n \in \mathbb{Z}\}$ for some $\alpha, \beta \in \mathbb{C}$, termed generators.

2.2.2 Example. We will show that $I = \mathbb{Z} \cdot 7 + \mathbb{Z}(3 + \sqrt{-5})$ absorbs multiplication by $\mathbb{Z}[\sqrt{-5}]$. It suffices to check that $\sqrt{-5}I \subseteq I$, which we check on the generators:

$$\sqrt{-5} \cdot 7 = (-3) \cdot 7 + 7(3 + \sqrt{-5}) \in I$$
$$\sqrt{-5} \cdot (3 + \sqrt{-5}) = -5 + 3\sqrt{-5} = -2 \cdot 7 + 3(3 + \sqrt{-5}) \in I. \qquad \square$$

We're also interested in functions that respect the ring structure.

2.2.3 Definition. *Given rings R and S, a function $\varphi : R \to S$ is a **ring homomorphism** if it meets the following conditions for any $a, b \in R$:*

(a) $\varphi(a + b) = \varphi(a) + \varphi(b)$
(b) $\varphi(ab) = \varphi(a)\varphi(b)$
(c) $\varphi(1_R) = 1_S$

If φ is one-to-one and onto, it is called a **ring isomorphism**. If an isomorphism exists between two rings R and S, we say they are **isomorphic** and write $R \cong S$.

The notion of kernel links ideals and ring homomorphisms.

2.2.4 Definition. *The **kernel** of a ring homomorphism $\varphi : R \to S$ is*

$$\ker \varphi = \{a \in R : \varphi(a) = 0_S\}.$$

It's easy to check that $\ker \varphi$ is an ideal. It measures how far φ is from being injective, in the sense that φ is one-to-one if and only if $\ker \varphi = 0_R$. When φ is onto, we can reconstruct it from its kernel by means of the quotient ring construction.

Given a ring R and its ideal I, we define an equivalence relation $\sim$ on R by $a \sim b$ if $a - b \in I$. The equivalence class of $a \in R$ is the **coset** $a + I = \{a + x : x \in I\}$. The set of all cosets, i.e., the partition of R defined by $\sim$, is denoted by $R/I = \{a + I : a \in R\}$.

2.2.5 Proposition-Definition (Definition of a Quotient Ring). *Let I be an ideal of a ring R. The expressions*

$$(a + I) + (b + I) = (a + b) + I$$
$$(a + I)(b + I) = (ab) + I$$

are well-defined operations that make R/I into a ring.

The function $\pi : R \to R/I$ defined by $\pi(a) = a + I$ is a surjective ring homomorphism.

Proof. When we write an element of R/I as $a + I$, we are in fact choosing a representative of this coset, namely a. Any other $a' \in a + I$ would do, as $a' + I = a + I$. We need to check that the two operations, defined in terms of arbitrary coset representatives, in fact depend only on the cosets themselves.

We do this for multiplication, and leave the rest to you. If $a + I = a' + I$ and $b + I = b' + I$, we need to show that $ab + I = a'b' + I$. By the definition of cosets as equivalence classes, we have $a - a', b - b' \in I$. As I absorbs multiplication, we get $ab - a'b' = a(b - b') + b'(a - a') \in I$. ∎

2.2.6 Theorem (First Isomorphism Theorem for Rings). *Let $\varphi : R \to S$ be a ring homomorphism. The assignment $a + \ker \varphi \mapsto \varphi(a)$ gives a well-defined ring isomorphism $\tilde{\varphi} : R/\ker \varphi \to \varphi(R)$ that satisfies $\varphi = \pi \circ \tilde{\varphi}$. We say that $\tilde{\varphi}$ is* **induced** *from φ.*

2.2.7 Example. Take $R = \mathbb{Z}$ and $I = 5\mathbb{Z}$. We usually think of $\mathbb{Z}/5\mathbb{Z}$ as the set $\{0, 1, 2, 3, 4\}$. To reduce $a \in \mathbb{Z}$ modulo 5, we find $5k \in 5\mathbb{Z}$ for which $a + 5k \in \{0, 1, 2, 3, 4\}$. □

This generalizes to computing an arbitrary quotient R/I. Since $a + I = a + x + I$ for all $x \in I$, we look for an $x \in I$ that makes $a + x$ as simple as possible. We often refer to $a + I$ as "a modulo I."

2.2.8 Example. It's not hard to check that $I = \mathbb{Z} \cdot 11 + \mathbb{Z}(4 - \sqrt{5})$ is an ideal of the ring $R = \mathbb{Z}[\sqrt{5}]$. Fix $a = k + l\sqrt{5} \in R$, and look for $x = m \cdot 11 + n(4 - \sqrt{5}) \in I$ for which

$$a + x = (k + 11m + 4n) + (l - n)\sqrt{5}$$

is as simple as possible. A natural choice would be to take $n = l$, so that $a + x$ is in fact in $\mathbb{Z}$. That integer changes by a multiple of 11 as we vary m. We pick

m so that $k+4l+11m \in \{0,1,2,\ldots,10\}$. Thus, each coset has a representative in $\{0,1,2,\ldots,10\}$, which strongly suggests (but doesn't prove!) that $R/I \cong \mathbb{Z}/11\mathbb{Z}$.

For a rigorous proof we use Thm. 2.2.6 and look for a surjective ring homomorphism $\varphi : R \to \mathbb{Z}/11\mathbb{Z}$ with kernel I. Any such homomorphism satisfies $\varphi(\sqrt{5})^2 = \varphi(5) = 5 \pmod{11}$. Since $4^2 \equiv 5 \pmod{11}$, we can easily check that $\varphi(x+y\sqrt{5}) = x+4y \pmod{11}$ is the desired ring homomorphism.

To illustrate, let's reduce $\alpha = 2 + 3\sqrt{5}$ modulo I. Here $k = 2, l = 3$, so $n = 3$ and $m = -1$: $(2 + 3\sqrt{5}) + ((-1) \cdot 11 + 3(4 - \sqrt{5})) = 3$. Thus, $\alpha \equiv 3 \pmod{I}$, which we also see from $\varphi(2 + 3\sqrt{5}) = 2 + 3 \cdot 4 \equiv 3 \pmod{11}$. $\square$

2.2.9 Example. Now take $R = \mathbb{Z}[i], I = 3\mathbb{Z}[i] = \mathbb{Z} \cdot 3 + \mathbb{Z} \cdot 3i$. Take $a = k + li \in R$ and look for $x = 3m + 3ni \in I$ which makes

$$a + x = (k + 3m) + (l + 3n)i$$

as simple as possible. We can't always kill the imaginary part, but we can make it $0, 1$, or 2. We similarly adjust the real part to get

$$R/I \cong \{k + li : k, l \in \{0,1,2\}\},$$

at least as sets. In fact, $R/I \cong \mathbb{Z}/3\mathbb{Z} \times \mathbb{Z}/3\mathbb{Z}$, the isomorphism being induced, as in Thm. 2.2.6, by the homomorphism

$$\varphi : R \to \mathbb{Z}/3\mathbb{Z} \times \mathbb{Z}/3\mathbb{Z}, \quad \varphi(k + li) = (k \bmod 3, l \bmod 3). \qquad \square$$

In both examples, we're really only determining the additive group structure of the quotient R/I. Its multiplication is determined by the multiplication in R, as in Prop.-Def. 2.2.5. We will develop a systematic method for computing similar quotients in Sec. 3.4 .

Exercises

2.2.1. Let $X \subseteq R$ be an arbitrary subset of a ring R. Prove that the smallest ideal containing X, called the ideal **generated** by X, is given by

$$\langle X \rangle = \left\{ \sum_{i=1}^{k} r_i x_i : k \in \mathbb{N}, r_i \in R, x_i \in X \right\}.$$

2.2.2. Let $R = \mathbb{Z}[\sqrt{19}]$ and put $I_b = \mathbb{Z} \cdot 5 + \mathbb{Z}(b + \sqrt{19})$ for $b \in \{0,1,2,3,4\}$. Which of the five I_b are ideals of R?

2.2.3. Find all values of $a \in \mathbb{N}$ for which $\mathbb{Z} \cdot a + \mathbb{Z}(9 + \sqrt{26})$ is an ideal of the ring $R = \mathbb{Z}[\sqrt{26}]$.

2.2.4. Let $\varphi : R \to S$ be a ring homomorphism. Show φ is injective if and only if $\ker \varphi = 0$.

2.2.5. Let R be a ring. Show that there is exactly one ring homomorphism $\varphi : \mathbb{Z} \to R$, and that char R (see Exer. 2.1.4) is the non-negative generator of the ideal ker φ.

2.2.6. Let R be a ring with p elements, for p prime. Show that R is isomorphic to $\mathbb{Z}/p\mathbb{Z}$.

2.2.7. Let R be a ring of prime characteristic p. Show that the function from R to itself given by $a \mapsto a^p$ is a ring homomorphism, called the **Frobenius endomorphism**.

2.2.8. Prove Prop.-Def. 2.2.5 and Thm. 2.2.6.

2.2.9. Let $\sigma : R \to S$ be a *surjective* ring homomorphism, and let I be an ideal of R. Show that the assignment $a + I \mapsto (\sigma a) + \sigma(I)$ gives a well-defined ring homomorphism $\sigma \mod I : R/I \to S/\sigma(I)$.

2.2.10. Let I be an ideal of a ring R, and let $\pi : R \to R/I$ be the ring homomorphism from Prop.-Def. 2.2.5. Prove that the assignment $J \mapsto \pi^{-1}J$ gives a bijection between the set of ideals of R/I and the set of ideals of R containing I.

2.2.11. For both examples Ex. 2.2.8 and Ex. 2.2.9, check that φ is an onto ring homomorphism with kernel I. Find all other such homomorphisms.

2.2.12. In Ex. 2.2.9, we determined the additive group structure of $\mathbb{Z}[i]/3\mathbb{Z}[i]$. Write down its multiplication table, as determined by Prop.-Def. 2.2.5. (Since we know multiplication is commutative, you only need to fill in half the table.) What are the units in $\mathbb{Z}[i]/3\mathbb{Z}[i]$?

2.2.13. Given a ring R and a subset $I \subseteq R$ below, prove that I is an ideal. Then imitate Ex. 2.2.8 to guess the quotient R/I, prove your guess is correct using the First Isomorphism Theorem 2.2.6, and finally reduce the given α modulo I:

(a) $R = \mathbb{Z}[\sqrt{22}]$, $I = \mathbb{Z} \cdot 7 + \mathbb{Z}(1 + \sqrt{22})$, $\alpha = 5 - 4\sqrt{22}$
(b) $R = \mathbb{Z}[\sqrt{46}]$, $I = \mathbb{Z} \cdot 23 + \mathbb{Z}\sqrt{46}$, $\alpha = -15 + 29\sqrt{46}$

2.2.14. This is the first in a series of exercises that show how ideals resolve the failure of unique factorization of 6 in $\mathbb{Z}[\sqrt{-5}]$, Ex. 1.5.4, along the lines envisaged by Kummer.

(a) Show that the following sets are ideals of $\mathbb{Z}[\sqrt{-5}]$:

$$P_1 = \mathbb{Z} \cdot 2 + \mathbb{Z}(1 + \sqrt{-5}),$$
$$P_2 = \mathbb{Z} \cdot 3 + \mathbb{Z}(1 + \sqrt{-5}),$$
$$P_3 = \mathbb{Z} \cdot 3 + \mathbb{Z}(-1 + \sqrt{-5}).$$

(b) For any $X \subseteq \mathbb{C}$, we put $\bar{X} = \{\bar{z} : z \in X\}$. Show that $\bar{P}_1 = P_1$ and $\bar{P}_2 = P_3$.
(c) Compute the quotient $\mathbb{Z}[\sqrt{-5}]/P_i$ for $i = 1, 2, 3$. (Continued in Exer. 2.3.6.)

2.2.15. Let F be a field. List all ideals of F. Deduce that any ring homomorphism from F to a ring R is injective.

2.2.16. Let K be a subfield of L.

(a) Prove that L is a K-vector space.
(b) Let $f \colon L \to L$ be a ring homomorphism. Prove that f is a K-linear transformation if and only if $f(k) = k$ for all $k \in K$.
(c) Assume $\dim_K L < \infty$. Let $f \colon L \to L$ be a ring homomorphism fixing K pointwise as in part (b). Prove f is bijective, and therefore is a ring isomorphism.

2.2.17. Let F be a field with finitely many elements. By Exer. 2.1.4(b) char $F = p$, a positive prime. Exer. 2.2.5 then defines a ring homomorphism $\mathbb{Z}/p\mathbb{Z} \hookrightarrow F$. Deduce from this, using Exer. 2.2.16, that F has p^n elements, for some n.

2.3 Principal Ideals

Divisibility in a general ring R is just as interesting as in $\mathbb{Z}$. For $a, b \in R$, we say that a **divides** b, written $a \mid b$, if $b = ac$ for some $c \in R$. If an ideal I of R contains a, then by the absorption property it must contain all elements divisible by a. In other words, $Ra \subseteq I$, where

$$Ra = \{ra : r \in R\}.$$

It's easy to check that Ra is itself an ideal, hence the smallest ideal containing a.

2.3.1 Definition. *An ideal I of R is a **principal ideal** if $I = Ra$ for some $a \in R$. Any such a is called a **generator** of I.*

If the ring R is fixed, we sometimes write $\langle a \rangle$ for Ra, in keeping with Exer. 2.2.1. When $R = \mathbb{Z}$ and the generator is a specific number, we follow tradition and write the generator on the left, e.g., $3\mathbb{Z}$ instead of $\mathbb{Z}3$.

2.3.2 Example. Let R be the ring $\mathbb{Z}[\sqrt{-5}]$. We claim that its ideal $I = \mathbb{Z} \cdot 7 + \mathbb{Z}(3 + \sqrt{-5})$ from Ex. 2.2.2 is not principal. If I had a generator α, we could find $\gamma, \delta \in R$ such that

$$7 = \gamma\alpha \text{ and } 3 + \sqrt{-5} = \delta\alpha.$$

Taking the norm, we get two multiplicative identities in $\mathbb{Z}$,

$$49 = \mathrm{N}\gamma \cdot \mathrm{N}\alpha \text{ and } 14 = \mathrm{N}\delta \cdot \mathrm{N}\alpha,$$

which imply $\mathrm{N}\alpha \mid \gcd(49, 14) = 7$. The equation $\mathrm{N}(x + y\sqrt{-5}) = x^2 + 5y^2 = 7$ has no solution in $\mathbb{Z}$, so $\mathrm{N}\alpha = 1$, α is a unit, and $I = R\alpha = R$. To arrive at a contradiction, it suffices to show that $1 \notin I$. This is because the equation $1 = a \cdot 7 + b(3 + \sqrt{-5})$ has for its only solution $a = 1/7 \notin \mathbb{Z}$ and $b = 0$. □

The notion of a principal ideal is most useful when dealing with rings similar to $\mathbb{Z}$.

2.3.3 Definition. *A ring $\mathcal{D}$ is an* **integral domain** *if it has no zero-divisors: for all $a, b \in \mathcal{D}$ with $ab = 0$, we have $a = 0$ or $b = 0$.*

Integral domains are precisely the rings in which **cancellation** holds: for all $a, b, c \in \mathcal{D}$ with $a \neq 0$, $ab = ac$ implies $b = c$. Any subring of a field F is an integral domain: the equality $ab = ac$ remains valid in F, where we can cancel a by multiplying both sides by $a^{-1} \in F$. For an example of a ring that isn't an integral domain, consider $\mathbb{Z}/6\mathbb{Z}$: in it, $2 \cdot 3 = 0$, while $2 \neq 0$ and $3 \neq 0$.

You can easily check the following basic properties of principal ideals.

2.3.4 Proposition. *Let R be a ring and $a, b \in R$.*

(a) $Ra = R$ if and only if $a \in R^{\times}$.
(b) The following three statements are equivalent: (i) $a \mid b$; (ii) $b \in Ra$; (iii) $Rb \subseteq Ra$.
(c) Assume that R is an integral domain. Then $Ra = Rb$ if and only if $a = bu$, for some $u \in R^{\times}$.

2.3.5 Definition. *A* **principal ideal domain (PID)** *is an integral domain in which every ideal is principal.*

The prototypical PID is $\mathbb{Z}$. Other familiar PIDs include any field F, as well as its polynomial ring $F[x]$. In fact, all three satisfy the stronger property of having a version of the division algorithm.

2.3.6 Definition. *Let $\mathcal{D}$ be an integral domain. A* **Euclid size** *on $\mathcal{D}$ is a function $\nu : \mathcal{D} \setminus 0 \to \mathbb{Z}_{\geq 0}$ with the following property: for any $a, b \in \mathcal{D}, b \neq 0$, there exist $q, r \in \mathcal{D}$ such that $a = bq + r$, and either $r = 0$ or $\nu(r) < \nu(b)$. The integral domain $\mathcal{D}$ is called a* **Euclid domain** *if there exists a Euclid size on it.*[1]

2.3.7 Example. Here are the Euclid sizes for the three families of PIDs mentioned before Def. 2.3.6:

(a) $\nu : \mathbb{Z} \setminus 0 \to \mathbb{Z}_{\geq 0}, \quad \nu(n) = |n|$
(b) $\nu : F \setminus 0 \to \mathbb{Z}_{\geq 0}, \quad \nu(x) = 1$
(c) $\nu : F[x] \setminus 0 \to \mathbb{Z}_{\geq 0}, \quad \nu(f(x)) = \deg f(x)$

In the last example, the degree of the zero polynomial is not defined, which is why the general definition excludes 0 from the domain of ν. □

---

[1] This is slightly nonstandard terminology. A Euclid size is commonly termed a "Euclidean norm," and a ring equipped with one a "Euclidean domain" We prefer the term "Euclid size" to avoid confusion with the field norm $N\alpha = \alpha\bar{\alpha}$.

The argument of Lemma 1.2.11 for $\mathbb{Z}[i]$ generalizes to show that an arbitrary Euclid domain is a PID. Indeed, the proofs of unique factorization in $\mathbb{Z}$, $\mathbb{Z}[i]$ and $\mathbb{Z}[\omega]$ follow the same outline:

Division algorithm $\Rightarrow$ All ideals are principal $\Rightarrow$ Euclid's algorithm
$\Rightarrow$ Euclid's lemma $\Rightarrow$ Unique factorization.

This train of thought proves unique factorization in both a Euclid domain and a PID; the Euclid domain merely boards it at the first stop (the division algorithm), the PID at the second. Since there are many more PIDs than Euclid domains, we will organize our study of arithmetic around the question, when is the ring of integers in a quadratic field a PID? For an example of a PID that can't be equipped with a Euclid size, see Exer. 2.3.9.

Exercises

2.3.1. Show $\mathcal{D}$ is an integral domain if and only if $ab = ac$ implies $b = c$ for all $a, b, c \in \mathcal{D}, a \neq 0$.

2.3.2. Prove Prop. 2.3.4

2.3.3.* For a nonsquare $D \in \mathbb{Z}$, prove that $\mathbb{Q}[\sqrt{D}] \cong \mathbb{Q}[x]/\langle x^2 - D \rangle$.

2.3.4. Put $\delta = \frac{1+\sqrt{-23}}{2}$ and $R = \mathbb{Z} + \mathbb{Z}\delta$.

(a) Prove that R is a ring.
(b) Prove that $I = \mathbb{Z} \cdot 3 + \mathbb{Z}(1 - \delta)$ is an ideal of R.
(c) Prove that I isn't principal.

2.3.5. Let $R = \mathbb{Z}[\sqrt{2}]$. Find $\alpha, \beta \in R$ for which $\langle 3 - 7\sqrt{2} \rangle = \mathbb{Z}\alpha + \mathbb{Z}\beta$.

2.3.6. Prove that the ideals P_1, P_2, P_3 from Exer. 2.2.14 are all nonprincipal. (Continued in Exer. 2.4.7.)

2.3.7. Show that the three functions of Ex. 2.3.7 are indeed Euclid size functions on their respective rings.

2.3.8. Let $\mathcal{D}$ be a Euclid domain with size function $\nu : \mathcal{D} \setminus 0 \to \mathbb{Z}_{\geq 0}$.

(a) One often additionally requires that $\nu(a) \leq \nu(ab)$ for all $a, b \in \mathcal{D} \setminus 0$, and calls such a ν a **strong Euclid size**. If $\mathcal{D}$ is a Euclid domain with size ν, show that the formula

$$\nu'(a) = \min\{\nu(ax) : x \in \mathcal{D} \setminus 0\}$$

defines a strong Euclid size on $\mathcal{D}$. We may therefore assume that every Euclid domain has a strong Euclid size without losing generality.

(b) Under that assumption, put $m = \min\{\nu(a) : a \in \mathcal{D} \setminus 0\}$. Describe the set of elements of smallest size, $\{u \in \mathcal{D} \setminus 0 : \nu(u) = m\}$.

2.3.9.* In this exercise, we show that $\mathcal{D} = \mathbb{Z}[\frac{1+\sqrt{-19}}{2}]$ is not a Euclid domain. We argue by contradiction, assuming that there is a strong Euclid size ν on $\mathcal{D}$.

(a) Find $\mathcal{D}^\times$.
(b) Show that 2 and 3 are irreducible in $\mathcal{D}$.
(c) Let $a \in \mathcal{D}$ have the second-smallest size, i.e., next-smallest after the m of Exer. 2.3.8(b). What does Exer. 2.3.8(b) tell you about the size of $\mathcal{D}/\mathcal{D}a$?
(d) Deduce a contradiction to (b) by showing that $a \mid 2$ or $a \mid 3$.

We will see in Ex. 5.4.3 that $\mathbb{Z}[\frac{1+\sqrt{-19}}{2}]$ is a PID.

2.3.10. Let K be an arbitrary field. Use the division algorithm on $K[x]$ to show that a polynomial of degree d in $K[x]$ can have at most d roots in K.

2.3.11. Let K be a finite field. Since $K^\times$ is a finite abelian group, it is isomorphic to a product of cyclic groups,

$$(2.3.8) \qquad K^\times \cong \mathbb{Z}/d_1\mathbb{Z} \times \mathbb{Z}/d_2\mathbb{Z} \times \cdots \times \mathbb{Z}/d_r\mathbb{Z},$$

where the $d_i \in \mathbb{N}$ satisfy $d_r \mid d_{r-1} \mid \cdots \mid d_1$. We will show that $K^\times$ is cyclic by showing that $d_1 = |K^\times|$, and therefore $i = 1$.

Assume that $d_1 < |K^\times|$. Then by the divisibility condition after (2.3.8), we have that $x^{d_1} = 1$ for all $x \in K^\times$. Show that this contradicts Exer. 2.3.10

2.4 Operations on Ideals

We will extend various multiplicative notions from ring elements to ideals, much as Kummer envisaged for his ideal numbers. Our first definition is motivated by Prop. 2.3.4(b).

2.4.1 Definition. *Let I and J be ideals of a ring R. If $J \subseteq I$, we say that I* **divides** *J and write $I \mid J$. In the special case when $J = Ra$ is principal, we write $I \mid a$ for the equivalent statements $I \mid Ra$ and $a \in I$.*

"Divide and con(tain)" is a mnemonic to help you remember that $I \mid J$ is simply an alternative notation for $I \supseteq J$. The redundancy allows us to systematically translate statements about divisibility of numbers into conjectures about containment of ideals, which are usually easier to prove.

2.4.2 Example. Consider the transitivity of divisibility in $\mathbb{Z}$:

$$(2.4.3) \qquad \text{If } a \mid b \text{ and } b \mid c, \text{then } a \mid c.$$

A plausible generalization to ideals I, J, K in a ring R would be:

$$(2.4.4) \qquad \text{If } I \mid J \text{ and } J \mid K, \text{then } I \mid K.$$

This statement is merely *analogous* to the true statement (2.4.3). To prove it, we replace | with $\supseteq$, which reduces (2.4.4) to the usual transitivity of inclusion

$$\text{If } I \supseteq J \text{ and } J \supseteq K, \text{then } I \supseteq K.$$

For more examples, see Exer. 2.4.5.   □

To extend the analogy between ideals and numbers, we define some multiplication-related operations on ideals in a ring R.

2.4.5 Definition. *For two ideals I and J of R, we define the following sets, each of which is itself an ideal:*

$$I + J = \{x + y : x \in I, y \in J\}$$
$$I \cap J = \{x : x \in I \text{ and } x \in J\}$$
$$IJ = \{\textstyle\sum_{i=1}^{m} x_i y_i : x_i \in I, y_i \in J\}.$$

Remarks:

(a) The ideal $I + J$ is the smallest ideal containing both I and J. In the language of divisibility, it is the biggest ideal dividing both. We therefore think of $I + J$ as the greatest common divisor of I and J (and *not* as the analogue of addition of numbers).

(b) In particular, if $I + J = R$, we say that I and J are **relatively prime ideals**. This happens if and only if there exist $x \in I$ and $y \in J$ with $x + y = 1$.

(c) Similarly, we think of $I \cap J$ as the least common multiple of I and J.

(d) The definition of IJ is the least transparent of the lot. You might be tempted to define the product of I and J as the *set* of all products xy, where $x \in I$ and $y \in J$. Alas, this set is not closed under addition. As we defined it, IJ is the smallest ideal containing all products xy as x ranges over I and y over J.

(e) The definition of IJ extends multiplication of ring elements to ideals, in the sense that $(Ra)(Rb) = R(ab)$.

2.4.6 Example. Let $R = \mathbb{Z}[\sqrt{19}]$. Check that the following are ideals in R:

$$I = \mathbb{Z} \cdot 2 + \mathbb{Z}(1 + \sqrt{19}), \quad J = \mathbb{Z} \cdot 3 + \mathbb{Z}(2 + \sqrt{19}).$$

By definition, IJ contains the four products of a generator of I with a generator of J:

$$2 \cdot 3 = 6$$
$$2 \cdot (2 + \sqrt{19}) = 4 + 2\sqrt{19}$$
(2.4.7)
$$(1 + \sqrt{19}) \cdot 3 = 3 + 3\sqrt{19}$$
$$(1 + \sqrt{19})(2 + \sqrt{19}) = 21 + 3\sqrt{19}.$$

A generic product $(r\cdot2+s(1+\sqrt{19}))(t\cdot3+u(2+\sqrt{19}))$ is a $\mathbb{Z}$-linear combination of the four elements above:

$$IJ = \{k \cdot 6 + l(3 + 3\sqrt{19}) + m(4 + 2\sqrt{19}) + n(21 + 3\sqrt{19}) : k, l, m, n \in \mathbb{Z}\}.$$

We would like to write IJ in the form $IJ = \mathbb{Z}a + \mathbb{Z}(b + \sqrt{19})$. To this end, we look for relations among the generators listed in (2.4.7).

We observe that $-1 + \sqrt{19} = (3 + 3\sqrt{19}) - (4 + 2\sqrt{19}) \in IJ$, and that the last three generators in (2.4.7) are $\mathbb{Z}$-linear combinations of 6 and $-1 + \sqrt{19}$:

$$3 + 3\sqrt{19} = 6 + 3(-1 + \sqrt{19})$$
$$4 + 2\sqrt{19} = 6 + 2(-1 + \sqrt{19})$$
$$21 + 3\sqrt{19} = 4 \cdot 6 + 3(-1 + \sqrt{19}).$$

We conclude that $IJ = \mathbb{Z} \cdot 6 + \mathbb{Z}(-1 + \sqrt{19})$. Similarly, one finds that $I + J = \{k \cdot 2 + l(1 + \sqrt{19}) + m \cdot 3 + n(2 + \sqrt{19}) : k, l, m, n \in \mathbb{Z}\}$. In particular, $I + J$ contains $3 - 2 = 1$, so that $I + J = R$ and I and J are relatively prime. □

The procedure for finding relations among the generators in the example may seem ad hoc, but it is clearly pure linear algebra, which we will study in depth in Ch. 3.

Many results about division and congruences in $\mathbb{Z}$ turn out to be special cases of ideal-theoretic propositions. As an example, the Chinese remainder theorem in $\mathbb{Z}$ follows from the following general result.

2.4.8 Proposition (The Chinese Remainder Theorem). *Let I and J be two relatively prime ideals of a ring R. Then*

$$R/IJ \cong R/I \times R/J.$$

The ring structure on the product is given by componentwise addition and multiplication.

Proof. Our only general tool for proving that two rings are isomorphic is the First Isomorphism Theorem (Thm. 2.2.6). It will imply the proposition once we construct a surjective homomorphism $\varphi\colon R \to R/I \times R/J$ whose kernel is precisely IJ. The natural candidate is the homomorphism $\varphi(a) = (a + I, a + J)$.

In a much-used move, we deduce from $I + J = R$ that there exist $x \in I$ and $y \in J$ such that $x + y = 1$. Pick any $a, b \in R$. As $bx + ay = bx + a(1 - x) = b(1 - y) + ay$, we find that

$$\varphi(bx + ay) = (a + x(b - a) + I, b + y(a - b) + J) = (a + I, b + J),$$

and φ is indeed onto.

Observe that $\varphi(a) = (a + I, a + J) = (I, J)$, the zero element in $R/I \times R/J$, if and only if $a \in I$ and $a \in J$. In brief, $\ker \varphi = I \cap J$, so the proposition will be proved once we show that $I \cap J = IJ$. The inclusion $\supseteq$ holds in general. As for $\subseteq$, take any $a \in I \cap J$. Using the x and y as above, we see that

$$a = a \cdot 1 = a(x + y) = ax + ay \in IJ. \qquad \blacksquare$$

2.4.9 Example. Let's take $R = \mathbb{Z}$, $I = 12\mathbb{Z}$, and $J = 17\mathbb{Z}$. The two ideals are relatively prime, since

$$(2.4.10) \qquad 1 = \gcd(12, 17) = (-7) \cdot 12 + 5 \cdot 17 \in I + J.$$

The elementary version of the Chinese Remainder Theorem has two parts. The first part asserts that for any $r, s \in \mathbb{Z}$, there exists an $a \in \mathbb{Z}$ satisfying both congruences

$$(2.4.11) \qquad a \equiv r \pmod{12} \text{ and } a \equiv s \pmod{17}.$$

That is just the restatement of the surjectivity of the homomorphism $\varphi(a) = (a \mod 12, a \mod 17)$ of Prop. 2.4.8. The second part, which claims that a is unique modulo $12 \cdot 17$, is equivalent to the injectivity of φ.

Solutions to the two particular systems,

$$a_{10} \equiv 1 \pmod{12}, \quad a_{10} \equiv 0 \pmod{17}, \text{ and}$$
$$a_{01} \equiv 0 \pmod{12}, \quad a_{01} \equiv 1 \pmod{17},$$

can be read off from the result of Euclid's Algorithm in (2.4.10): $a_{10} = 5 \cdot 17$, $a_{01} = -7 \cdot 12$. The solution to an arbitrary system of congruences (2.4.11) is then a linear combination of these two particular solutions: $a = ra_{10} + sa_{01} = 85r - 84s$. $\qquad \square$

Exercises

2.4.1. Check that the three operations of Def. 2.4.5 indeed produce ideals.

2.4.2. (a) Let I and J be ideals of a ring R. Show that $I \cup J$ is an ideal of R if and only if either $I \subseteq J$ or $J \subseteq I$.

(b) Let $I_1 \subseteq I_2 \subseteq \cdots$ be an ascending chain (finite or infinite) of ideals in R. Prove that $\cup_{n \geq 1} I_n$ is an ideal.

2.4.3. Let I, J, and K be ideals of a ring R. Prove the following identities:

(a) $I + J = J + I, IJ = JI$
(b) $(I + J) + K = I + (J + K), (IJ)K = I(JK)$
(c) $(I + J)K = IK + JK$
(d) $IR = RI = I$

(e) If $IJ = R$, then $I = J = R$. In other words, R is the only ideal with an inverse for ideal multiplication.

(f) If $I \subseteq J$, then $IK \subseteq JK$.

2.4.4. An ideal I of a ring R is **finitely generated** if $I = \langle X \rangle$ for a finite subset $X = \{x_1, \ldots, x_n\} \subseteq R$ (for notation, see Exer. 2.2.1). Prove that $I = Rx_1 + \cdots + Rx_n$.

2.4.5. Here is a list of true statements about the integers:

(a) $1 \mid a \mid 0$
(b) If $c \mid a$ and $c \mid b$, then $c \mid \gcd(a, b)$.
(c) If $a \mid c$ and $b \mid c$, then $\mathrm{lcm}(a, b) \mid c$.
(d) $a \mid b$ if and only if $b = ac$ for some $c \in \mathbb{Z}$.
(e) $ab = \mathrm{lcm}(a, b) \gcd(a, b)$
(f) A chain of divisors in $\mathbb{Z}$ is a sequence $a_1, a_2, \ldots \in \mathbb{Z}$ satisfying $a_2 \mid a_1, a_3 \mid a_2$, etc. Any chain of divisors stabilizes: there is an $n_0 \in \mathbb{N}$ such that $a_{n+1} = a_n$ for $n \geq n_0$.

Following Ex. 2.4.2, translate these statements into conjectures about ideals in a ring R, and state them in terms of both ideal divisibility and containment. If you like a challenge, decide which of the resulting conjectures are true (and give a proof), and which aren't (and give a counterexample).

2.4.6. Consider the ring $R = \mathbb{Z}[\sqrt{7}]$ and its two subsets

$$I = \mathbb{Z} \cdot 6 + \mathbb{Z}(2 + 2\sqrt{7}), \quad J = \mathbb{Z} \cdot 21 + \mathbb{Z}(7 + \sqrt{7}).$$

Check that both are ideals. Write the ideals $I + J$, IJ and $I \cap J$ as lattices of the form $\mathbb{Z}a + \mathbb{Z}(b + c\sqrt{7})$.

2.4.7. Recall the ideals P_1, P_2 and P_3 from Exer. 2.2.14. Denote by $\langle \alpha \rangle$ the principal ideal generated by $\alpha \in \mathbb{Z}[\sqrt{-5}]$. In the following diagram, check the vertical equalities by performing the indicated ideal multiplications:

$$\langle 6 \rangle = \langle 2 \rangle \cdot \langle 3 \rangle = \langle 1 + \sqrt{-5} \rangle \cdot \langle 1 - \sqrt{-5} \rangle$$
$$\| \qquad \| \qquad\qquad \| \qquad\qquad \|$$
$$P_1^2 \cdot P_2 P_3 = \quad P_1 P_2 \quad \cdot \quad P_1 P_3.$$

These identities almost resolve the apparent failure of unique factorization of 6 in $\mathbb{Z}[\sqrt{-5}]$, following Kummer's program of Sec. 1.5. All that is left is to show that P_1, P_2, P_3 are "prime," which awaits in Exer. 2.5.5.

2.4.8. Let I, J, and K be ideals in a ring R. Assume that $I + J = R$, $I \mid K$ and $J \mid K$. Prove that $IJ \mid K$.

2.5 Prime and Maximal Ideals

Before we generalize unique factorization to ideals, we need to decide which among them are the right analog of prime numbers.

2.5.1 Definition. *An ideal M of R is* **maximal** *if $M \neq R$ and R is the only ideal strictly containing M.*

In other words, if I is an ideal with $M \subseteq I \subseteq R$, then either $I = M$ or $I = R$.

2.5.2 Definition. *An ideal P of R is* **prime** *if $P \neq R$ and if, for all $a, b \in R$, $ab \in P$ implies $a \in P$ or $b \in P$.*

In terms of ideal divisibility (Def. 2.4.1), an ideal $P \neq R$ is prime if, for all $a, b \in R$,

$$P \mid ab \text{ implies } P \mid a \text{ or } P \mid b.$$

Prime ideals thus satisfy a generalization (*à la* Kummer) of Euclid's lemma. By definition, R itself is neither a prime nor a maximal ideal.

2.5.3 Example. Let $p \in \mathbb{N}$ be prime. The following chain of implications shows that $\mathbb{Z}p$ is a prime ideal of $\mathbb{Z}$:

$$ab \in \mathbb{Z}p \; \Rightarrow \; p \mid ab \; \Rightarrow \; p \mid a \text{ or } p \mid b \; \Rightarrow \; a \in \mathbb{Z}p \text{ or } b \in \mathbb{Z}p.$$

In fact, $\mathbb{Z}p$ is also maximal. Suppose that an ideal $\mathbb{Z}a$ satisfies $\mathbb{Z}p \subsetneq \mathbb{Z}a$, or, in elementary terms, $a \mid p$ and $a \neq p$. Since p is prime, a must be ± 1, so that $\mathbb{Z}a = \mathbb{Z}$, as required by Def. 2.5.1.

On the other hand, $\mathbb{Z} \cdot 6$ is not a prime ideal of $\mathbb{Z}$: $2 \cdot 3$ is in $\mathbb{Z} \cdot 6$, but neither 2 nor 3 is. $\square$

The following pair of propositions characterizes maximal and prime ideals in terms of their quotients.

2.5.4 Proposition. *An ideal $P \subsetneq R$ is prime if and only if R/P is an integral domain.*

Proof. In the diagram

$$(a + P)(b + P) = 0 + P \Rightarrow (a + P = 0 + P \text{ or } b + P = 0 + P)$$
$$\Updownarrow \qquad\qquad\qquad\qquad \Updownarrow \qquad\qquad\qquad \Updownarrow$$
$$ab \in P \qquad\qquad \Rightarrow \qquad (a \in P \quad\text{ or }\quad b \in P)$$

the top row is the statement that R/P is an integral domain, while the bottom row asserts that P is a prime ideal. The two statements are equivalent since their constituents are, by the definition of the quotient ring R/P. ∎

2.5.5 Proposition. *An ideal $M \subsetneq R$ is maximal if and only if R/M is a field.*

Proof. Let $M \subsetneq R$ be a maximal ideal and $a + M \in R/M$ a nonzero element. This means that $a \notin M$, so that we have a strict inclusion $M \subsetneq M + Ra$. The maximality of M implies $M + Ra = R$, giving us an $m \in M$ and $b \in R$ such that $m + ba = 1$. Then

$$(b + M)(a + M) = ba + M = (1 - m) + M = 1 + M,$$

and $a + M$ has a multiplicative inverse, $b + M$. For the converse, simply reverse the argument.   ∎

2.5.6 Corollary. *Any maximal ideal is a prime ideal.*

Proof. If M is a maximal ideal of R, then R/M is a field. In particular, R/M is an integral domain, so that M is a prime ideal.   ∎

Conversely, when is a prime ideal maximal? The corollary to the following proposition gives a sufficient condition that will be satisfied in rings of quadratic integers.

2.5.7 Proposition. *Let $\mathcal{D}$ be a ring with finitely many elements. Then $\mathcal{D}$ is an integral domain if and only if $\mathcal{D}$ is a field.*

Proof. Assume that $\mathcal{D}$ is an integral domain and take $a \in \mathcal{D} \setminus 0$. Consider the homomorphism of additive groups $\mu_a : (\mathcal{D}, +) \to (\mathcal{D}, +)$ defined by $\mu_a(x) = ax$. As $\mathcal{D}$ is an integral domain, we find that

$$\ker \mu_a = \{x \in \mathcal{D} : ax = 0\} = 0.$$

This means that μ_a is injective and therefore also surjective, since $\mathcal{D}$ is finite. In particular, there exists a $b \in \mathcal{D}$ such that $\mu_a(b) = ab = 1$. Thus, each nonzero element of $\mathcal{D}$ is invertible, which makes $\mathcal{D}$ a field. The converse is trivial, as every field is an integral domain.   ∎

2.5.8 Corollary. *Let P be a prime ideal of a ring R. If R/P is finite, then P is a maximal ideal.*

As suggested in Ch. 1, we are shifting the focus of arithmetic from elements to ideals. It is nevertheless interesting that a direct generalization of prime numbers to an arbitrary integral domain yields a somewhat more general notion than the requirement that Euclid's lemma be satisfied.

2.5.9 Definition. *Let $\mathcal{D}$ be an integral domain and $p \in \mathcal{D} \setminus \mathcal{D}^\times$.*

(a) *p is **irreducible** if, for all $a, b \in R$, $p = ab$ implies either $a \in \mathcal{D}^\times$ or $b \in \mathcal{D}^\times$.*

(b) *p is a **prime element** if, for all $a, b \in R$, $p \mid ab$ implies $p \mid a$ or $p \mid b$.*

Check the following basic consequences of the definition.

2.5.10 Proposition. *Let $\mathcal{D}$ be an integral domain.*

(a) An element $p \in \mathcal{D}$ is prime if and only if $\mathcal{D}p$ is a prime ideal.
(b) Any prime element of $\mathcal{D}$ is irreducible.

The converse of (b) is not generally true.

2.5.11 Example. We saw in Ex. 1.5.4 that 3 is irreducible in $\mathbb{Z}[\sqrt{-5}]$. It is, however, not a prime element of $\mathbb{Z}[\sqrt{-5}]$: $3 \mid (1 + \sqrt{-5})(1 - \sqrt{-5})$, but 3 does not divide either factor. □

Exercises

2.5.1.* Let P be an ideal of a ring R. Show that P is a prime ideal if and only if the following analog of Euclid's lemma holds for any two ideals I and J of R:
$$P \mid IJ \text{ if and only if } P \mid I \text{ or } P \mid J.$$

2.5.2. Prove Prop. 2.5.10.

2.5.3. Prove that a ring R is an integral domain if and only if $\{0\}$ is a prime ideal of R.

2.5.4. In each example, decide whether I is a prime ideal of R. If it isn't, find a prime ideal $P \supset I$:

(a) $R = \mathbb{Z}[\sqrt{14}], \quad I = \mathbb{Z} \cdot 35 + \mathbb{Z}(7 + \sqrt{14})$
(b) $R = \mathbb{Z}[\sqrt{-33}], \quad I = \mathbb{Z} \cdot 7 + \mathbb{Z}(4 + \sqrt{-33})$
(c) $R = \mathbb{Z}[\frac{1+\sqrt{21}}{2}], \quad I = \mathbb{Z} \cdot 85 + \mathbb{Z}(9 + \frac{1+\sqrt{21}}{2})$
(d) $R = \mathbb{Z}[\sqrt{35}], \quad I = \mathbb{Z} \cdot 1 + \mathbb{Z}(-5 + \sqrt{35})$

2.5.5. To conclude the series of exercises the factorization of 6 in $\mathbb{Z}[\sqrt{-5}]$, show that the ideals P_1, P_2, P_3 of Exer. 2.2.14 are all prime. Use your calculation from part (c) of that exercise.

2.5.6. Let R be a PID. Show that each nonzero prime ideal P of R is maximal.

2.5.7. Let $R = \mathbb{C}[x, y]$, the ring of polynomials in two variables with coefficients in $\mathbb{C}$. Show that $Rx + Ry$ is a maximal ideal, and that Rx is a prime ideal that isn't maximal.

2.5.8.* Prove that there are infinitely many prime ideals in $\mathbb{Z}[\sqrt{D}]$, for any $D \in \mathbb{Z}$ which isn't a complete square.

2.5.9. Let M be a maximal ideal of a ring R.

(a) Show that M is the *only* maximal ideal of R if and only if $R^\times = R \setminus M$. In that case, R is called a **local ring**.
(b) Show that R/M^n is a local ring with a unique maximal ideal that is M/M^n.

2.5.10. Let R be a local ring with maximal ideal M.

(a) Check that $1 + M^n$ is a subgroup of $R^\times$.
(b) Show that reduction modulo M defines a surjective group homomorphism $R^\times \to (R/M)^\times$ with kernel $1 + M$.
(c) Prove that $1 + m + M^{n+1} \mapsto m + M^{n+1}$ is an isomorphism between the multiplicative group $(1 + M^n)/(1 + M^{n+1})$ and the additive group M^n/M^{n+1}. Show that the latter group has the structure of an R/M-vector space.

Chapter 3
Lattices

3.1 Group Structure of Lattices

Most of the rings and ideals we saw in Ch. 1 are lattices, such as $\mathbb{Z} + \mathbb{Z}\sqrt{-5}$ and $\mathbb{Z} \cdot 7 + \mathbb{Z}(3 + \sqrt{-5})$. We can roughly think of them as "vector spaces with coefficients in $\mathbb{Z}$," and explore them with the tools of linear algebra. For this we'll need a simple bit of terminology. Let V be a complex vector space and R any subring of $\mathbb{C}$. A linear combination $a_1 v_1 + \cdots + a_n v_n \in V$ with coefficients $a_i \in R$ is said to be R-**linear**, or **defined over** R. Similar terminology applies to other linear algebra constructs: we talk of R-linear transformations, linear (in)dependence over R, etc.

Crucial to the proofs of the division algorithm in $\mathbb{Z}[i]$ and $\mathbb{Z}[\omega]$ was the geometry of lattices as subsets of a **plane**, a two-dimensional real vector space such as $\mathbb{R}^2$ or $\mathbb{C}$.

3.1.1 Definition. *Let V be a plane. A* **lattice** *$\Lambda \subset V$ is an additive subgroup of V of the form $\Lambda = \mathbb{Z}v_1 + \mathbb{Z}v_2 = \{av_1 + bv_2 : a, b \in \mathbb{Z}\}$, where the vectors $v_1, v_2 \in V$ are linearly independent over $\mathbb{R}$, i.e., noncollinear. Any such pair $\{v_1, v_2\}$ is called a* **basis** *of Λ.*

To shed light on this definition, we give two examples. The first is a lattice; the second is not.

3.1.2 Example. Write Λ_0 (resp. V_0) for the group (resp. $\mathbb{R}$-vector space) of 2×1 column vectors with entries in $\mathbb{Z}$ (resp. $\mathbb{R}$). We take $\{[\begin{smallmatrix}1\\0\end{smallmatrix}], [\begin{smallmatrix}0\\1\end{smallmatrix}]\}$ as a basis for both. We will treat Λ_0 as our "bureau of standards" lattice. For any lattice $\Lambda = \mathbb{Z}v_1 + \mathbb{Z}v_2$ inside a plane V, the assignment

$$[\begin{smallmatrix}1\\0\end{smallmatrix}] \mapsto v_1, \quad [\begin{smallmatrix}0\\1\end{smallmatrix}] \mapsto v_2$$

defines an $\mathbb{R}$-linear isomorphism $V_0 \xrightarrow{\sim} V$ whose restriction to Λ_0 is an isomorphism of abelian groups $\Lambda_0 \Lambda$. □

M. Trifković, *Algebraic Theory of Quadratic Numbers*, Universitext, DOI 10.1007/978-1-4614-7717-4_3, © Springer Science+Business Media New York 2013

3.1.3 Example. Consider the group

$$L = \mathbb{Z}\begin{bmatrix} 1 \\ 0 \end{bmatrix} + \mathbb{Z}\begin{bmatrix} \sqrt{2} \\ 0 \end{bmatrix} \subset \mathbb{R}^2.$$

As a group, L is isomorphic to $\mathbb{Z}^2$. Indeed, since $\sqrt{2}$ is irrational, there is no linear relation $a\begin{bmatrix} 1 \\ 0 \end{bmatrix} + b\begin{bmatrix} \sqrt{2} \\ 0 \end{bmatrix} = \begin{bmatrix} 0 \\ 0 \end{bmatrix}$ with $a, b \in \mathbb{Z}$. However, L is not a lattice in the sense of Def. 3.1.1 because $\begin{bmatrix} 1 \\ 0 \end{bmatrix}$ and $\begin{bmatrix} \sqrt{2} \\ 0 \end{bmatrix}$ are linearly dependent over $\mathbb{R}$: $\sqrt{2}\begin{bmatrix} 1 \\ 0 \end{bmatrix} - \begin{bmatrix} \sqrt{2} \\ 0 \end{bmatrix} = \begin{bmatrix} 0 \\ 0 \end{bmatrix}$. Besides, L looks wrong: it's contained in the x-axis, and it's not discrete. □

3.1.4 Definition. *Let $\Lambda = \mathbb{Z}v_1 + \mathbb{Z}v_2 \subset V$ be a lattice. The **fundamental parallelogram** for Λ with respect to basis $\{v_1, v_2\}$ is the set*

$$\Pi_{\{v_1,v_2\}} = \{t_1 v_1 + t_2 v_2 : t_1, t_2 \in [0,1)\} \subset V.$$

The linear independence of v_1 and v_2 means that the plane is tiled by the translates $v + \Pi_{\{v_1,v_2\}}$, as v ranges over Λ (see Figs. 1.2 and 1.3). That is the content of the following proposition, whose proof is left to you. How is the peculiar interval $[0,1)$ in the definition of $\Pi_{\{v_1,v_2\}}$ relevant to the uniqueness?

3.1.5 Proposition. *For any $w \in V$ there exists a unique pair $a, b \in \mathbb{Z}$ such that $w - (av_1 + bv_2)$ lies in the fundamental parallelogram $\Pi_{\{v_1,v_2\}}$.*

Example 3.1.2 reduces the study of the group structure of any lattice to that of the "bureau of standards" lattice Λ_0.

3.1.6 Theorem. *If $A \subseteq \Lambda_0$ is a subgroup, then A is one of the following:*

(a) 0;
(b) $\mathbb{Z}t$, for some $t \in \Lambda_0$; or,
*(c) $\mathbb{Z}t_1 + \mathbb{Z}t_2$, for two linearly independent vectors $t_1, t_2 \in \Lambda_0$. Such a subgroup is termed a **sublattice**.*

The term "linearly independent" in part (c) is ambiguous. Does it mean "over $\mathbb{Z}$" or "over $\mathbb{R}$"? Ex. 3.1.3 cautions that, in general, two vectors linearly independent over $\mathbb{Z}$ may no longer be so over $\mathbb{R}$. For the specific lattice $\Lambda_0 \subset V_0$, however, the two interpretations are equivalent.

3.1.7 Lemma. *If $t_1, t_2 \in \Lambda_0$ and $t_2 = \alpha t_1, \alpha \in \mathbb{R}$, then there t_1 and t_2 are linearly dependent over $\mathbb{Z}$.*

Proof. Let $t_1 = \begin{bmatrix} r_1 \\ s_1 \end{bmatrix}$ and $t_2 = \begin{bmatrix} r_2 \\ s_2 \end{bmatrix} = \alpha\begin{bmatrix} r_1 \\ s_1 \end{bmatrix}$. We may as well assume that $r_1 \neq 0$, so that $\alpha = r_2/r_1$, and

$$r_2 t_1 - r_1 t_2 = r_2 t_1 - r_1\left(\frac{r_2}{r_1}\right) t_1 = 0. \qquad \blacksquare$$

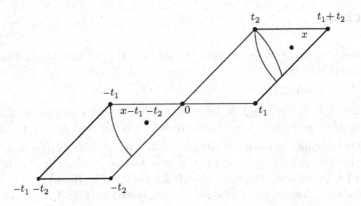

Fig. 3.1 The fundamental parallelogram for basis t_1 and t_2 covering a third vector x.

Proof of Thm. 3.1.6. Assume that $A \neq 0$. Define the length-squared function $\lambda : \Lambda_0 \to \mathbb{Z}_{\geq 0}$ by $\lambda\begin{bmatrix} a \\ b \end{bmatrix} = a^2 + b^2$. Choose a shortest nonzero $t_1 \in A$, that is, one where λ attains its minimum on $A \setminus 0$. If $A = \mathbb{Z}t_1$, we are in case (b).

Assume that $A \neq \mathbb{Z}t_1$. Let t_2 be one of the elements of minimal length in $A \setminus \mathbb{Z}t_1$. We claim that t_1 and t_2 are linearly independent over $\mathbb{Z}$ (and therefore over $\mathbb{R}$, by Lemma 3.1.7). If not, there exist $a, b \in \mathbb{Z} \setminus 0$ such that $at_1 = bt_2$. The division algorithm produces $q, r \in \mathbb{Z}$ such that $a = qb + r$ and $0 \leq r < |b|$.

Put $t' = (r/b)t_1 = (r/a)t_2 = t_2 - qt_1$, which is in A since A is a group. We have that $\lambda t' = (r/b)^2 \lambda t_1 < \lambda t_1$, which contradicts the choice of t_1 as the shortest nonzero element in A, unless $t' = t_2 - qt_1 = 0$. This in turn contradicts $t_2 \notin \mathbb{Z}t_1$, and shows that t_1 and t_2 are linearly independent.

Suppose there existed $x \in A \setminus (\mathbb{Z}t_1 + \mathbb{Z}t_2)$. After possibly translating by a $\mathbb{Z}$-linear combination of t_1 and t_2, we may assume that x is inside the fundamental parallelogram $\Pi_{\{t_1, t_2\}}$. As $\lambda x \geq \lambda t_2$ by choice of t_2, x lies on or outside the circle of radius $\sqrt{\lambda t_2}$ centered at 0. Then, as it lies inside the fundamental parallelogram, it is at a distance less than $\sqrt{\lambda t_1}$ from $t_1 + t_2$ (see Fig. 3.1). This implies that $\lambda(x - t_1 - t_2) < \lambda t_1$, contradicting $x - t_1 - t_2 \in A \setminus 0$. ∎

The statement of Thm. 3.1.6 is purely about the group structure of Λ_0. The proof, by contrast, makes full use of the geometry of Λ_0 as a subset of the plane. Thm. 3.1.6 is our first taste of the difference between linear algebra over $\mathbb{Z}$, and the usual $\mathbb{R}$-linear sort. The only two-dimensional subvector space of a plane is the plane itself. A lattice, by contrast, has infinitely many sublattices.

Exercises

3.1.1. For a fixed $j \in \mathbb{Z}$, prove that the set $\Lambda = \{[\begin{smallmatrix}x\\y\end{smallmatrix}] \in \Lambda_0 : x \equiv jy \pmod{p}\}$ is a sublattice of Λ_0.

3.1.2. Prove Prop. 3.1.5

3.1.3. Let $\Lambda = \mathbb{Z}v_1 + \mathbb{Z}v_2 \subset V$ be a lattice. Show that every basis of Λ is of the form $\{kv_1 + lv_2, mv_1 + nv_2\}$, where $k, l, m, n \in \mathbb{Z}$ and $kn - lm = \pm 1$.

3.1.4. One should generally distrust proofs by picture. Give a rigorous (e.g., trigonometric) argument for the next-to-last sentence in the proof of Thm. 3.1.6 by showing that the two circles centered at 0 and $t_1 + t_2$ indeed cover the fundamental parallelogram, as suggested by Fig. 3.1.

3.1.5. Show that the area of the fundamental parallelogram doesn't depend on the choice of basis.

3.1.6. Let Λ be a lattice and $\Lambda' \subseteq \Lambda$ a sublattice. Show that $|\Lambda/\Lambda'|$ is equal to the number of points of Λ inside a fundamental parallelogram of Λ' (with two of the edges excluded, complete with their endpoints, as in Def. 3.1.4). Conclude that $\{v_1, v_2\}$ is a basis of Λ if and only if the origin is the only point of Λ in the fundamental parallelogram Π_{v_1, v_2}.

3.2 Linear Algebra Over $\mathbb{Z}$

We now enlist column operations on matrices to systematically study sublattices of Λ_0. Unlike in standard undergraduate linear algebra, our matrices always have entries in $\mathbb{Z}$. Consequently, the only column (and later, row) operations we consider are the ones which don't introduce denominators.

For any ring R, we denote by $M_{k \times l}(R)$ the additive group of $k \times l$ matrices with entries in R. When $k = l$, the group $M_{k \times k}(R)$ is a noncommutative ring under matrix addition and multiplication. Its unit group, denoted $GL_k(R)$, consists of matrices with a two-sided inverse, i.e., those with determinant in $R^\times$.

We start with a simple criterion for a 2×2 matrix with real entries to actually have entries in $\mathbb{Z}$.

3.2.1 Proposition. *Let $\gamma \in M_{2 \times 2}(\mathbb{R})$. Then $\gamma \in M_{2 \times 2}(\mathbb{Z})$ if and only if $\gamma \Lambda_0 \subseteq \Lambda_0$.*

Proof. Let $\gamma = [\begin{smallmatrix}a&b\\c&d\end{smallmatrix}]$. As $\Lambda_0 = \mathbb{Z}[\begin{smallmatrix}1\\0\end{smallmatrix}] + \mathbb{Z}[\begin{smallmatrix}0\\1\end{smallmatrix}]$, we find that $\gamma \Lambda_0 = \mathbb{Z}[\begin{smallmatrix}a\\c\end{smallmatrix}] + \mathbb{Z}[\begin{smallmatrix}b\\d\end{smallmatrix}]$, which proves the Proposition. ■

3.2.2 Corollary. *Let $\gamma \in M_{2 \times 2}(\mathbb{R})$. Then $\gamma \Lambda_0 = \Lambda_0$ if and only if*

$$\gamma \in GL_2(\mathbb{Z}) = \left\{ [\begin{smallmatrix}a&b\\c&d\end{smallmatrix}] : a, b, c, d \in \mathbb{Z} \text{ and } ad - bc = \pm 1 \right\}$$

Proof. If $\gamma\Lambda_0 = \Lambda_0$, then $\gamma V_0 = V_0$, which implies that γ is invertible in $M_{2\times 2}(\mathbb{R})$. It then makes sense to replace the condition $\gamma\Lambda_0 = \Lambda_0$ by the equivalent one, $\gamma\Lambda_0 \subseteq \Lambda_0$ and $\gamma^{-1}\Lambda_0 \subseteq \Lambda_0$. Applying Prop. 3.2.1 to both inclusions, we see that $\gamma, \gamma^{-1} \in M_{2\times 2}(\mathbb{Z})$, so that $\gamma \in M_{2\times 2}(\mathbb{Z})^{\times} = GL_2(\mathbb{Z})$. The concrete description of $GL_2(\mathbb{Z})$ comes from Exer. 2.1.5 (b). ∎

3.2.3 Corollary. *A subgroup $\Lambda \subseteq \Lambda_0$ is a sublattice if and only if there exists a $\gamma \in M_{2\times 2}(\mathbb{Z})$ with $\Lambda = \gamma\Lambda_0$ and $\det\gamma \neq 0$.*

Proof. Assume that Λ is a sublattice of Λ_0. By Ex. 3.1.2, there exists a matrix $\gamma = \begin{bmatrix} a & b \\ c & d \end{bmatrix}$ defining an isomorphism $\gamma : V_0 \xrightarrow{\sim} V_0$ for which $\gamma\Lambda_0 = \Lambda$. Since $\Lambda \subseteq \Lambda_0$, Prop. 3.2.1 guarantees that $a, b, c, d \in \mathbb{Z}$, as desired. As $\begin{bmatrix} a \\ c \end{bmatrix}$ and $\begin{bmatrix} b \\ d \end{bmatrix}$ are linearly independent over $\mathbb{Z}$, Lemma 3.1.7 implies that they remain independent over $\mathbb{R}$. We then know from linear algebra over $\mathbb{R}$ that $\det\gamma \neq 0$. The converse is similar. ∎

If $\Lambda = \gamma'\Lambda_0 = \gamma\Lambda_0$, then $\gamma^{-1}\gamma'\Lambda_0 = \Lambda_0$, so that $\alpha = \gamma^{-1}\gamma' \in GL_2(\mathbb{Z})$ by the corollary 3.2.2. To easily compute with Λ, we'd like $\gamma' = \gamma\alpha$ to be as close as possible to a diagonal matrix.

Finding such a γ' is no harder for a general $\gamma \in M_{k\times l}(\mathbb{Z})$. For any $\alpha \in GL_l(\mathbb{Z})$, we can compute $\gamma\alpha$ by applying to γ a sequence of integral column operations (see Exer. 3.2.2). They come in three flavors, each illustrated in the last column by its effect on the matrix $\begin{bmatrix} 1 & 0 \\ 0 & 1 \end{bmatrix}$:

Operation	Notation	Example
Add $a \cdot$ col. i to col. j	$C_j + aC_i$	$\begin{bmatrix} 1 & a \\ 0 & 1 \end{bmatrix}$ $(i=1, j=2)$
Switch columns	$C_i \circlearrowright C_j$	$\begin{bmatrix} 0 & 1 \\ 1 & 0 \end{bmatrix}$
Multiply col. i by -1	$-C_i$	$\begin{bmatrix} -1 & 0 \\ 0 & 1 \end{bmatrix}$ $(i=1)$.

This is familiar from linear algebra over $\mathbb{R}$, with the additional requirement that all the constants, like a in the first flavor, must be in $\mathbb{Z}$. Over $\mathbb{R}$, we can multiply a column by any nonzero real number, in other words a unit in $\mathbb{R}$. Over $\mathbb{Z}$ the multiplier has to be in $\mathbb{Z}^{\times} = \{\pm 1\}$. This leaves multiplication by -1 as the only nontrivial column operation of the third flavor.

3.2.4 Example. Here's an example of simplifying a matrix using column operations. The sequence of operations is secretly motivated by Thm. 3.2.6.

$$\begin{bmatrix} 14 & 4 \\ 25 & 10 \end{bmatrix} \xrightarrow{C_1 - 2C_1} \begin{bmatrix} 6 & 4 \\ 5 & 10 \end{bmatrix} \xrightarrow{C_1 \circlearrowright C_2} \begin{bmatrix} 4 & 6 \\ 10 & 5 \end{bmatrix} \xrightarrow{C_1 - 2C_2}$$

$$\begin{bmatrix} -8 & 6 \\ 0 & 5 \end{bmatrix} \xrightarrow{-C_1} \begin{bmatrix} 8 & 6 \\ 0 & 5 \end{bmatrix} \xrightarrow{C_2 - C_1} \begin{bmatrix} 8 & -2 \\ 0 & 5 \end{bmatrix}$$

Any further column operation that preserves the upper triangular form has to be $C_2 + kC_1$, for some $k \in \mathbb{Z}$. This makes the upper right-hand entry $-2 + 8k$, which is never zero. We can't diagonalize the original matrix using only integral column operations. □

Integral column operations at best reduce a matrix to an essentially upper triangular one.

3.2.5 Definition. *A* $k \times k$ *matrix* $A = [a_{ij}]$ *is* **upper triangular** *if all entries below the main diagonal are* 0, *i.e.,* $a_{ij} = 0$ *for* $i > j$.

A $k \times l$ *matrix is in* **column-reduced form** *if it has an upper triangular matrix of maximal size,* $\min(k,l) \times \min(k,l)$, *in its lower right corner, and zeros in any remaining columns.*

Explicitly, a column-reduced matrix has one of the following forms: ∇ if $k = l$, $[0|\nabla]$ if $k < l$, $\left[\frac{*}{\nabla}\right]$ if $k > l$. In all three cases, ∇ is an upper triangular matrix, 0 stands for any number of zero columns, and $*$ stands for any number of arbitrary rows.

3.2.6 Theorem (Integral Column Reduction). *Any matrix with entries in* $\mathbb{Z}$ *can be put into column-reduced form by a series of integral column operations.*

Proof. The following algorithm specifies the promised column operations.

COLUMN REDUCTION ALGORITHM

1. INPUT: $k \times l$ matrix X. Put $A := X$, which means that variable A gets the value of variable X.

2. Switch columns of A until all zero columns are on the left, to get a matrix of the form $A := [0|B]$.

3. If the last row of A is zero, go to Step 5. Otherwise, find the entry of smallest nonzero absolute value in the last row, and put it in the bottom right position by switching columns. This gives

$$A := \begin{bmatrix} & * & \\ a_1 & \cdots & a_l \end{bmatrix},$$

where $a_l \neq 0$, and either $a_i = 0$ or $|a_i| \geq |a_l|$ for all $i < l$.

4. Write $a_i = q_i a_l + r_i, 0 \leq r_i < |a_l|$ for $1 \leq i < l$. Perform the following $l - 1$ column operations:

$$A = \begin{bmatrix} & & & * & & \\ q_1 a_l + r_1 & q_2 a_l + r_2 & \cdots & q_{r-1} a_l + r_{l-1} & a_l \end{bmatrix} \xrightarrow{C_i - q_i C_l}$$

$$\begin{bmatrix} & & *' & & \\ r_1 & r_2 & \cdots & r_{l-1} & a_l \end{bmatrix} = C.$$

Put $A := C$.

5. If A consists of a single row $[0 \cdots 0 \; a_l]$ (or, when $l = 1$, $[a_1]$), OUTPUT: A.

6. Else, if A has more than two rows, the last of which is $[0 \ldots 0 \; a_l]$, perform the operation is of the form

$$A = \begin{bmatrix} D & * \\ 0 \cdots 0 & a_l \end{bmatrix} \rightarrow \begin{bmatrix} D' & *' \\ 0 \cdots 0 & a_l \end{bmatrix} = E,$$

where D' is obtained by recursively applying this algorithm to $X := D$. OUTPUT: E.

7. Else, if neither of the conditions of Steps 5 and 6 applies, go to Step 2 with the current value of A.

Step 2 enforces the requirement on zero columns of Def. 3.2.5. If some $r_i \neq 0$, Step 4 decreases the smallest absolute value of a nonzero entry in the last row. This must eventually stop and produce a matrix A that satisfies the condition of Step 5 or Step 6. The latter step builds up, row by row, an upper triangular matrix in the bottom right corner. It does so by shrinking the matrix that we are column-reducing. This reduction must eventually bottom out with a $1 \times l$ matrix as in Step 5. The procedure is thus guaranteed to terminate, and deserves the title "algorithm." ∎

3.2.7 Definition. *Let $A \in M_{k \times l}(\mathbb{Z})$. Define* col A, *the* **column group of** A, *as the additive group of all $\mathbb{Z}$-linear combinations of columns of A.*

We're interested in col A because it is invariant under column operations.

3.2.8 Proposition. *Let A' be a matrix obtained from a matrix A by a series of integral column operations. Then* col $A =$ col A'.

Proof. All column operations involve permuting columns of A or taking their $\mathbb{Z}$-linear combinations. The group col A is closed under both, so col $A' \subseteq$ col A. The reverse inclusion holds because every column operation can be undone by another of the same type. ∎

Exercises

3.2.1. Perform column reduction on the following matrices. First, apply the Column Reduction Algorithm verbatim; then do it again with the shortest sequence of operations that you can find.

(a) $\begin{bmatrix} 2 & 0 & 7 \\ -3 & 0 & 5 \end{bmatrix}$, (b) $\begin{bmatrix} -8 & 2 & 0 \\ 3 & -6 & 3 \\ 10 & -5 & 10 \\ 2 & 2 & -2 \end{bmatrix}$, (c) $\begin{bmatrix} 1 & 2 & 3 \\ 4 & 5 & 6 \\ 7 & 8 & 9 \end{bmatrix}$.

3.2.2. Let $\alpha \in \mathrm{GL}_l(\mathbb{Z})$. Show that there is a sequence of integral column operations that takes the $l \times l$ identity matrix to α. Prove that the same sequence takes a matrix $\gamma \in M_{k \times l}(\mathbb{Z})$ to $\gamma\alpha$.

3.2.3. Are $\begin{bmatrix} 7 & 0 \\ -1 & 3 \end{bmatrix}$ and $\begin{bmatrix} 6 & 2 \\ 10 & 5 \end{bmatrix}$ related by integral column operations?

3.2.4.* Check the first claim in the proof of Cor. 3.2.2: for any $\gamma \in M_{2\times 2}(\mathbb{R})$, $\gamma\Lambda_0 = \Lambda_0$ implies $\gamma V_0 = V_0$.

3.2.5. Take $A = [a_{ij}] \in M_{k \times l}(\mathbb{Z})$.

(a) Let $I_A = \langle a_{11}, a_{12}, \dots, a_{kl} \rangle$ be the ideal in $\mathbb{Z}$ generated by all the entries of A. Show that applying integral column (or row) operations to A doesn't change I_A.
(b) Column-reducing the $1 \times n$ matrix $[\,a_1 \ a_2 \ \cdots \ a_n\,]$ yields a matrix of the form $[\,0 \ 0 \ \cdots \ 0 \ x\,]$. Use part (a) to determine x without actually performing any column operations.

3.2.6. Suppose that a square matrix $A \in M_{k \times k}(\mathbb{Z})$ can be column-reduced to an upper triangular matrix T. Prove that, up to sign, every diagonal entry of T depends only on A.

3.3 Computing with Ideals

Column reduction is an essential tool for calculating with ideals, as we'll show through examples in $\mathbb{Z}[\sqrt{-5}]$. The additive group isomorphism $\mathbb{Z}[\sqrt{-5}] \cong \Lambda_0$ given by $a + b\sqrt{-5} \leftrightarrow \begin{bmatrix} a \\ b \end{bmatrix}$ identifies an ideal I with a subgroup $\Lambda_I \subseteq \Lambda_0$. In the examples below, it will be obvious that Λ_I is a sublattice of Λ_0. In general, we will prove that in Ch. 4.

3.3.1 Example. Find the product of ideals

$$I = \mathbb{Z} \cdot 3 + \mathbb{Z}(1 - \sqrt{-5}) \text{ and } J = \mathbb{Z} \cdot 7 + \mathbb{Z}(3 + \sqrt{-5}).$$

As in Ex. 2.4.6, the ideal IJ consists of $\mathbb{Z}$-linear combinations of $3 \cdot 7$, $3(3 + \sqrt{-5})$, $7(1 - \sqrt{-5})$, and $(1 - \sqrt{-5})(3 + \sqrt{-5}) = 8 - 2\sqrt{-5}$. The lattice Λ_{IJ} is then the column group of the matrix A below, and doesn't change when we apply the Column Reduction Algorithm:

$$A = \begin{bmatrix} 21 & 9 & 7 & 8 \\ 0 & 3 & -7 & -2 \end{bmatrix} \xrightarrow[C_2+C_4]{C_3-3C_4} \begin{bmatrix} 21 & 17 & -17 & 8 \\ 0 & 1 & -1 & -2 \end{bmatrix} \xrightarrow{C_2 \circlearrowright C_4}$$

$$\begin{bmatrix} 21 & 8 & -17 & 17 \\ 0 & -2 & -1 & 1 \end{bmatrix} \xrightarrow[C_2+2C_4]{C_3-C_4} \begin{bmatrix} 21 & 42 & -42 & 17 \\ 0 & 0 & 0 & 1 \end{bmatrix} \xrightarrow[C_3+2C_1]{C_2-2C_1}$$

$$\begin{bmatrix} 21 & 0 & 0 & 17 \\ 0 & 0 & 0 & 1 \end{bmatrix} \xrightarrow[C_1 \circlearrowright C_3]{C_4-C_1} \begin{bmatrix} 0 & 0 & 21 & -4 \\ 0 & 0 & 0 & 1 \end{bmatrix}$$

Beginning with the fourth arrow, we deviate from the algorithm and use a sequence of operations that, for this particular matrix, gets us to the reduced form faster. The column vectors $\begin{bmatrix} 21 \\ 0 \end{bmatrix}$ and $\begin{bmatrix} -4 \\ 1 \end{bmatrix}$ are clearly independent because of the zero in the first vector. The zero is there because the column-reduced form is essentially upper triangular.

Translating back from vectors in Λ_0 to elements of $\mathbb{Z}[\sqrt{-5}]$, we get that $IJ = \mathbb{Z} \cdot 21 + \mathbb{Z}(-4 + \sqrt{-5})$. The Column Reduction Algorithm doesn't care that the column group of the initial matrix came from an ideal. As a check on the computation, let's make sure that the group IJ indeed absorbs multiplication by $\mathbb{Z}[\sqrt{-5}]$. This we do by checking that $(\sqrt{-5})IJ \subseteq IJ$:

$$\sqrt{-5} \cdot 21 = 4 \cdot 21 + 21 \cdot (-4 + \sqrt{-5}) \in IJ$$
$$\sqrt{-5}(-4 + \sqrt{-5}) = -5 - 4\sqrt{-5} = (-1) \cdot 21 + (-4)(-4 + \sqrt{-5}) \in IJ. \quad \square$$

3.3.2 Example. Find $I + J$ for ideals

$$I = \mathbb{Z} \cdot 3 + \mathbb{Z}(1 + \sqrt{-5}) \text{ and } J = \mathbb{Z} \cdot 3 + \mathbb{Z}(1 - \sqrt{-5}).$$

By definition, $I + J$ is the smallest ideal containing both I and J. We get its generators by listing together the generators of I and J: $3, 1 + \sqrt{-5}, 3$ and $1 - \sqrt{-5}$. This list is obviously redundant, so we look for a basis of the column group of the initial matrix in the following column reduction:

$$\begin{bmatrix} 3 & 1 & 3 & 1 \\ 0 & 1 & 0 & -1 \end{bmatrix} \xrightarrow[C_1-C_3]{C_2+C_4} \begin{bmatrix} 0 & 2 & 3 & 1 \\ 0 & 0 & 0 & -1 \end{bmatrix} \xrightarrow[-C_4]{C_3-C_2}$$

$$\begin{bmatrix} 0 & 2 & 1 & -1 \\ 0 & 0 & 0 & 1 \end{bmatrix} \xrightarrow[C_4+C_3]{C_2-2C_3} \begin{bmatrix} 0 & 0 & 1 & 0 \\ 0 & 0 & 0 & 1 \end{bmatrix}$$

As $I + J = \mathbb{Z} \cdot 1 + \mathbb{Z}\sqrt{-5} = \mathbb{Z}[\sqrt{-5}]$, I and J are relatively prime. We may drop one of the 3's in the list of generators, so that this computation could just as well be done by reducing a 2×3 matrix. $\quad \square$

3.3.3 Example. Find $I \cap J$ for ideals

$$I = \mathbb{Z} \cdot 21 + \mathbb{Z}(4 + \sqrt{-5}) \text{ and } J = \mathbb{Z} \cdot 18 + \mathbb{Z}(4 + 2\sqrt{-5}).$$

If $\alpha \in I \cap J$, we can find $a, b, c, d \in \mathbb{Z}$ with $\alpha = a \cdot 21 + b(4 + \sqrt{-5}) = c \cdot 18 + d(4 + 2\sqrt{-5})$. Comparing coefficients, we get the following system of equations:

(3.3.4)
$$\begin{aligned} 21a + 4b - 18c - 4d &= 0 \\ b \qquad\quad\; -2d &= 0. \end{aligned}$$

The solution, with c and d as the free variables, is

$$a = \frac{2(9c - 2d)}{21} \text{ and } b = 2d.$$

We'd be done now if we were doing ordinary linear algebra, but we want more: a, b, c and d must be in $\mathbb{Z}$. This imposes the additional condition $2d \equiv 9c \pmod{21}$, which holds when $d = 15c + 21k$ for some $c, k \in \mathbb{Z}$. We can then write $\alpha \in I \cap J$ as

$$\alpha = c \cdot 18 + d(4 + 2\sqrt{-5}) = c \cdot (78 + 30\sqrt{-5}) + k \cdot (84 + 42\sqrt{-5}),$$

so $I \cap J = \mathbb{Z}(78 + 30\sqrt{-5}) + \mathbb{Z}(84 + 42\sqrt{-5})$. To simplify, we column-reduce the corresponding matrix:

$$\begin{bmatrix} 78 & 84 \\ 30 & 42 \end{bmatrix} \xrightarrow{C_2 - C_1} \begin{bmatrix} 78 & 6 \\ 30 & 12 \end{bmatrix} \xrightarrow{C_1 - 2C_2} \begin{bmatrix} 66 & 6 \\ 6 & 12 \end{bmatrix} \xrightarrow{C_2 - 2C_1}$$

$$\begin{bmatrix} 66 & -126 \\ 6 & 0 \end{bmatrix} \xrightarrow[-C_1]{C_1 \circlearrowleft C_2} \begin{bmatrix} 126 & 66 \\ 0 & 6 \end{bmatrix}$$

Finally, $I \cap J = \mathbb{Z} \cdot 126 + \mathbb{Z}(66 + 6\sqrt{-5})$. Check that this is indeed an ideal!   □

Exercises

3.3.1. Find $I + J, IJ$ and $I \cap J$ in each of the following examples. Check your answers by verifying that they are actually ideals, as in Ex. 3.3.1:

(a) $R = \mathbb{Z}[\sqrt{7}]$, $I = \mathbb{Z} \cdot 6 + \mathbb{Z}(2 + 2\sqrt{7})$, $J = \mathbb{Z} \cdot 21 + \mathbb{Z}(7 + \sqrt{7})$
(b) $R = \mathbb{Z}[\sqrt{-10}]$, $I = \mathbb{Z} \cdot 77 + \mathbb{Z}(-12 + \sqrt{-10})$, $J = \mathbb{Z} \cdot 91 + \mathbb{Z}(30 + \sqrt{-10})$
(c) $R = \mathbb{Z}[\frac{1+\sqrt{13}}{2}]$, $I = \mathbb{Z} \cdot 17 + \mathbb{Z}(-5 + \frac{1+\sqrt{13}}{2})$, $J = \mathbb{Z} \cdot 17 + \mathbb{Z}(4 + \frac{1+\sqrt{13}}{2})$
(d) $R = \mathbb{Z}[5i]$, $I = \mathbb{Z} \cdot 5 + \mathbb{Z} \cdot 5i$, $J = \mathbb{Z} \cdot 17 + \mathbb{Z}(3 + 5i)$
(e) $R = \mathbb{Z}[\sqrt{21}]$, $I = \mathbb{Z} \cdot 10 + \mathbb{Z}(-1 + \sqrt{21})$, $J = \mathbb{Z} \cdot 12 + \mathbb{Z}(-3 + \sqrt{21})$

3.4 Lattice Quotients

What about integral row operations? They too come in three flavors analogous to those for column operations on p. 49, and are denoted $R_j + aR_i$, $R_i \circlearrowright R_j$ and $-R_i$. Let $\alpha, \gamma \in M_{k \times k}(\mathbb{Z})$, with α invertible. The product $\gamma \alpha$ can be computed by a sequence of integral column operations. Similarly, a sequence of integral row operations computes the product $\alpha \gamma$. Row operations are less important to us because of the following asymmetry.

Each $\gamma \in M_{2 \times 2}(\mathbb{Z})$ with $\det \gamma \neq 0$ determines a sublattice $\Lambda = \gamma \Lambda_0$ of Λ_0. Then $(\gamma \alpha)\Lambda_0 = \gamma \Lambda_0 = \Lambda$, but $(\alpha \gamma)\Lambda_0 = \alpha \Lambda$. Multiplying γ by α on the right leaves Λ unchanged, whereas multiplying by α on the left replaces Λ by an isomorphic, but in general different, lattice $\alpha \Lambda$.

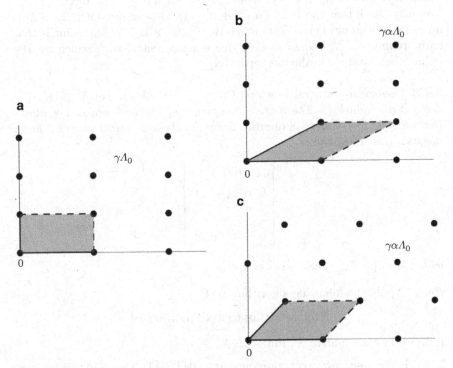

Fig. 3.2 Left vs. right multiplication by $\alpha \in \mathrm{GL}_2(\mathbb{Z})$.

3.4.1 Example. This is illustrated in Fig. 3.2 for $\gamma = \left[\begin{smallmatrix} 2 & 0 \\ 0 & 1 \end{smallmatrix}\right], \alpha = \left[\begin{smallmatrix} 1 & 1 \\ 0 & 1 \end{smallmatrix}\right]$. In each of the three pictures, the shaded parallelogram is the image of the standard fundamental parallelogram $\Pi_{\left[\begin{smallmatrix}1\\0\end{smallmatrix}\right],\left[\begin{smallmatrix}0\\1\end{smallmatrix}\right]}$ of Λ_0. Comparing pictures (a) and (b), we see that the set of bold dots hasn't changed, since $\gamma \alpha \Lambda_0 = \gamma \Lambda_0$. The points have, however, moved within it, as seen from the shear in the fundamental parallelogram. By contrast, in picture (c) the lattice $\alpha \gamma \Lambda_0$ itself is visibly different from $\gamma \Lambda_0$. $\square$

There is one kind of information about Λ that doesn't change when we replace it by $\alpha\Lambda$. For $v \in \Lambda_0$, the assignment $v + \Lambda \mapsto \alpha v + \alpha\Lambda$ gives a well-defined isomorphism $\Lambda_0/\Lambda \xrightarrow{\sim} \Lambda_0/\alpha\Lambda$. In other words, left multiplication by α doesn't change the structure of the quotient Λ_0/Λ. To understand quotients, then, we use both column and row operations to bring γ to the simplest possible form, a diagonal matrix.

3.4.2 Example. Let $\Lambda = \left[\begin{smallmatrix} 6 & 0 \\ 0 & 3 \end{smallmatrix}\right]\Lambda_0$, and take an arbitrary $\left[\begin{smallmatrix} x \\ y \end{smallmatrix}\right] \in \Lambda_0$. The division algorithm allows us to write $x = 6q + r$ and $y = 3s + t$, so that

$$\begin{bmatrix} x \\ y \end{bmatrix} + \Lambda = \begin{bmatrix} r \\ t \end{bmatrix} + q\begin{bmatrix} 6 \\ 0 \end{bmatrix} + s\begin{bmatrix} 0 \\ 3 \end{bmatrix} + \Lambda = \begin{bmatrix} r \\ t \end{bmatrix} + \Lambda.$$

Thus, any coset in Λ_0/Λ is represented by $\left[\begin{smallmatrix} r \\ t \end{smallmatrix}\right]$ with $0 \le r \le 5$ and $0 \le t \le 2$. You may check that the function $\left[\begin{smallmatrix} x \\ y \end{smallmatrix}\right] + \Lambda \mapsto (r, t) = (x \mod 6, y \mod 3)$ is an isomorphism of abelian groups $\Lambda_0/\Lambda \cong \mathbb{Z}/6\mathbb{Z} \times \mathbb{Z}/3\mathbb{Z}$. The point is that each column of $\left[\begin{smallmatrix} 6 & 0 \\ 0 & 3 \end{smallmatrix}\right]$ contains only one nonzero entry, with which we can adjust the x and y coordinates separately.   □

3.4.3 Theorem (Integral Row and Column Reduction). *Let $X \in M_{k \times l}(\mathbb{Z})$ and put $n = \min(k, l)$. There exists a sequence of row and column operations that takes X to a matrix in* **normal form**: *in the lower right corner it has a diagonal $n \times n$ submatrix*

$$\begin{bmatrix} d_n & 0 & \cdots & 0 \\ 0 & d_{n-1} & \cdots & 0 \\ \vdots & \vdots & \ddots & \vdots \\ 0 & \cdots & 0 & d_1 \end{bmatrix}$$

with $d_1 \mid d_2 \mid \cdots \mid d_n$, and zeros elsewhere.

Proof. Apply the following algorithm to A:

NORMAL REDUCTION ALGORITHM

1. INPUT: $k \times l$ matrix X. Put $A := X$.

2. If A has one row, column-reduce it. OUTPUT: the resulting matrix $\begin{bmatrix} 0 & \cdots & 0 & a_l \end{bmatrix}$.

3. Repeatedly apply the Column Reduction Algorithm of Thm. 3.2.6 and its analogue for row reduction, until A is of the form

$$A := \begin{bmatrix} B & 0 \\ 0 & a_{kl} \end{bmatrix}.$$

4. If $a_{kl} \nmid b_{ij}$ for some entry b_{ij} of B, perform the column operation

$$
A = \begin{bmatrix} & & 0 \\ & & \vdots \\ B & & 0 \\ & & \vdots \\ & & 0 \\ 0 & & a_{kl} \end{bmatrix} \xrightarrow{\;C_l + C_j\;} \begin{bmatrix} & & b_{1j} \\ & & \vdots \\ B & & b_{ij} \\ & & \vdots \\ & & b_{k-1,j} \\ 0 & & a_{kl} \end{bmatrix} = C,
$$

put $A := C$, and go to Step 3.

5. Else, if a_{kl} divides all the entries of B, perform the operation

$$
A = \begin{bmatrix} B & 0 \\ 0 & a_{kl} \end{bmatrix} \to \begin{bmatrix} B' & 0 \\ 0 & a_{kl} \end{bmatrix} = D,
$$

where B' is produced by recursively applying the algorithm to $X := B$.
OUTPUT: D.

If we are in Step 4, we can write $b_{ij} = q a_{kl} + r$ with $0 < r < |a_{kl}|$, since $a_{kl} \nmid b_{ij}$. In the following iteration of Step 3, we row-reduce to kill all the entries in the last column, bar the bottom one. The first step is

$$
\begin{bmatrix} & & * \\ & & \vdots \\ B & & b_{ij} \\ & & \vdots \\ & & * \\ 0 & & a_{kl} \end{bmatrix} \xrightarrow{\;R_i - q R_k\;} \begin{bmatrix} & & *' \\ & & \vdots \\ B & & r \\ & & \vdots \\ & & *' \\ 0 & & a_{kl} \end{bmatrix}.
$$

We've decreased the smallest nonzero absolute value of an entry in the last column, from $|a_{kl}|$ to at most r. That can repeat only finitely many times, after which we must end up in Step 5. This step recursively shrinks the matrix, eventually bottoming out in Step 2. The resulting matrix $B' = [b'_{ij}]$ is in normal form. Moreover, $a_{kl} \mid \gcd(b_{ij}) = \gcd(b'_{ij})$, which proves that divisibility condition on the diagonal entries of the output matrix D. ∎

3.4.4 Example. Let $\Lambda = \begin{bmatrix} -2 & 6 \\ 6 & 14 \end{bmatrix} \Lambda_0$. To find Λ_0 / Λ, we apply the Normal Reduction Algorithm:

$$
\begin{bmatrix} -2 & 6 \\ 6 & 14 \end{bmatrix} \xrightarrow[\;R_2 + 3R_1\;]{-C_1} \begin{bmatrix} 2 & 6 \\ 0 & 32 \end{bmatrix} \xrightarrow{\;C_2 - 3C_1\;} \begin{bmatrix} 2 & 0 \\ 0 & 32 \end{bmatrix} \xrightarrow[\;R_1 \circlearrowleft R_2\;]{C_1 \circlearrowleft C_2} \begin{bmatrix} 32 & 0 \\ 0 & 2 \end{bmatrix}.
$$

The row and column switching in the last operation was necessary to get the divisibility condition $d_1 \mid d_2$. We now find the quotient:

$$\Lambda_0/\Lambda \cong (\mathbb{Z}[\begin{smallmatrix}1\\0\end{smallmatrix}] + \mathbb{Z}[\begin{smallmatrix}0\\1\end{smallmatrix}])/(\mathbb{Z}[\begin{smallmatrix}32\\0\end{smallmatrix}] + \mathbb{Z}[\begin{smallmatrix}0\\2\end{smallmatrix}]) \cong \mathbb{Z}/32\mathbb{Z} \times \mathbb{Z}/2\mathbb{Z}.$$

Note that $|\Lambda_0/\Lambda| = 64 = |\det[\begin{smallmatrix}-2 & 6\\6 & 14\end{smallmatrix}]|.$ $\qquad\qquad\square$

3.4.5 Proposition. *Let $\Lambda = \gamma\Lambda_0$ for some $\gamma \in M_{2\times2}(\mathbb{Z})$ with $\det\gamma \neq 0$. The quotient Λ_0/Λ is finite and $|\Lambda_0/\Lambda| = |\det\gamma|$.*

Proof. Let $\alpha, \beta \in GL_2(\mathbb{Z})$ be the matrices corresponding to the sequences of row and column operations, respectively, which diagonalize γ as in Thm. 3.4.3:

$$\alpha\gamma\beta = \begin{bmatrix}d_2 & 0\\0 & d_1\end{bmatrix} \text{ with } d_1 \mid d_2.$$

Then

$$\Lambda_0/\Lambda = \Lambda_0/\gamma\Lambda_0 = \Lambda_0/(\alpha^{-1}[\begin{smallmatrix}d_2&0\\0&d_1\end{smallmatrix}]\beta^{-1})\Lambda_0$$
$$= \Lambda_0/(\alpha^{-1}[\begin{smallmatrix}d_2&0\\0&d_1\end{smallmatrix}]\Lambda_0) \cong \Lambda_0/[\begin{smallmatrix}d_2&0\\0&d_1\end{smallmatrix}]\Lambda_0 \cong \mathbb{Z}/d_2\mathbb{Z} \times \mathbb{Z}/d_1\mathbb{Z}.$$

Clearly $|\Lambda_0/\Lambda| = d_1 d_2$. On the other hand,

$$\det\gamma = (\det\alpha^{-1})\det[\begin{smallmatrix}d_2&0\\0&d_1\end{smallmatrix}](\det\beta^{-1}) = (\pm1)(d_1 d_2)(\pm1),$$

so that $|\Lambda_0/\Lambda| = |\det\gamma|$. $\qquad\qquad\blacksquare$

Exercises

3.4.1. Prove that the function $v + \Lambda \mapsto \alpha v + \alpha\Lambda$ is well-defined, and that it gives an isomorphism $\Lambda_0/\Lambda \xrightarrow{\sim} \Lambda_0/\alpha\Lambda$.

3.4.2. Verify that the function in Ex. 3.4.2 is a well-defined isomorphism.

3.4.3. In Thm. 3.4.3, describe d_1 in terms of the entries of the original matrix X.

3.4.4.* The Chinese Remainder Theorem says that if $\gcd(a,b) = 1$, then $\mathbb{Z}/a\mathbb{Z} \times \mathbb{Z}/b\mathbb{Z} \cong \mathbb{Z}/ab\mathbb{Z}$. Prove this by performing integral row and column reduction on a suitable matrix.

3.4.5. Consider the linear function $\mathbb{Z}^4 \to \mathbb{Z}^3$ defined by the matrix

$$A = \begin{bmatrix}2 & 5 & -1 & 2\\1 & 0 & 2 & -1\\-1 & 3 & 1 & 1\end{bmatrix}$$

Use integral row and column reduction to express the quotient $\mathbb{Z}^3/A\mathbb{Z}^4$ as a product of cyclic groups. (This is a concrete instance of the proof of one of the fundamental theorems of algebra: every finitely generated abelian group is a product of cyclic groups).

3.4.6. Let $\Lambda_1, \Lambda_2 \subseteq \Lambda$ be two sublattices with $|\Lambda/\Lambda_1|$ and $|\Lambda/\Lambda_2|$ relatively prime. Prove that $\Lambda = \Lambda_1 + \Lambda_2$.

Chapter 4
Arithmetic in $\mathbb{Q}[\sqrt{D}]$

4.1 Quadratic Fields

In Ch. 1, the examples of fields $\mathbb{Q}[i], \mathbb{Q}[\sqrt{-3}], \mathbb{Q}[\sqrt{-5}]$ and $\mathbb{Q}[\sqrt{319}]$ gave us a sense of the new phenomena that arise in the arithmetic of fields bigger than $\mathbb{Q}$: nonunique factorization, importance of ideals, infinite unit groups, etc. In this chapter we investigate those phenomena in a general quadratic field, using the algebraic toolkit from Chs. 2 and 3.

4.1.1 Definition. *A* **quadratic field** *is a field of the form*

$$\mathbb{Q}[\sqrt{D}] = \{a + b\sqrt{D} : a, b \in \mathbb{Q}\},$$

where $D \in \mathbb{Z}$ is not a perfect square. The field $\mathbb{Q}[\sqrt{D}]$ is called **imaginary** *if $D < 0$, and* **real** *if $D > 0$.*

All the theory we will develop applies to both real and imaginary quadratic fields, unless explicitly stated otherwise.

We say that D is square-free if it isn't divisible by any perfect square other than 1; equivalently, D is a product of distinct primes. When working in $\mathbb{Q}[\sqrt{D}]$, it is often useful to assume that D is square-free. This loses no generality: if $D' = n^2 D$, then $a + b\sqrt{D'} = a + bn\sqrt{D}$, so $\mathbb{Q}[\sqrt{D}] = \mathbb{Q}[\sqrt{D'}]$.

4.1.2 Definition. *Let $\alpha \in \mathbb{Q}[\sqrt{D}]$. The* **conjugate** *of $\alpha = a + b\sqrt{D}$ is $\bar{\alpha} = a - b\sqrt{D}$. We define the* **trace** *and the* **norm** *of α by*

$$\operatorname{Tr}\alpha = \alpha + \bar{\alpha}, \quad N\alpha = \alpha\bar{\alpha}.$$

Note that $\bar{\alpha}$ is independent of choice of $\sqrt{D}$. For $D < 0$, $\bar{\alpha}$ is the usual complex conjugate; for $D > 0$ we get a new function that doesn't extend to all of $\mathbb{R}$ (see Exer. 1.6.8). For $S \subseteq \mathbb{Q}[\sqrt{D}]$, we put as usual $\bar{S} = \{\bar{\alpha} : \alpha \in S\}$.

Let $\alpha \in \mathbb{Q}[\sqrt{D}] \setminus \mathbb{Q}$. The trace and the norm of α appear naturally when we look for polynomials $p(x) \in \mathbb{Q}[x]$ satisfying $p(\alpha) = 0$. By Exer. 1.3.1 we must also have $p(\bar{\alpha}) = 0$, so $p(x)$ must be divisible by

M. Trifković, *Algebraic Theory of Quadratic Numbers*, Universitext,
DOI 10.1007/978-1-4614-7717-4_4, © Springer Science+Business Media New York 2013

$$(x - \alpha)(x - \bar{\alpha}) = x^2 - (\alpha + \bar{\alpha})x + \alpha\bar{\alpha} = x^2 - (\mathrm{Tr}\,\alpha)x + \mathrm{N}\alpha.$$

The proof of the following basic properties is a good exercise.

4.1.3 Proposition. *Let* $\alpha \in \mathbb{Q}[\sqrt{D}]$.

(a) *For* $a, b \in \mathbb{Q}$, *we have*

$$\mathrm{Tr}(a + b\alpha) = 2a + (\mathrm{Tr}\,\alpha)b,$$
$$\mathrm{N}(a + b\alpha) = a^2 + (\mathrm{Tr}\,\alpha)ab + (\mathrm{N}\alpha)b^2.$$

(b) $\mathrm{Tr}\,\alpha$ *and* $\mathrm{N}\alpha$ *are rational. The functions* $\mathrm{Tr}\,:\,\mathbb{Q}[\sqrt{D}] \to \mathbb{Q}$ *and* $\mathrm{N}\,:\,\mathbb{Q}[\sqrt{D}]^\times \to \mathbb{Q}^\times$ *are homomorphisms of additive and multiplicative groups, respectively.*

(c) α *is a root of* $x^2 - (\mathrm{Tr}\,\alpha)x + \mathrm{N}\alpha$, *which is the unique such monic quadratic polynomial with rational coefficients, (unless* $\alpha \in \mathbb{Q}$).

(d) $\alpha \in \mathbb{Q}$ *if and only if* $\alpha = \bar{\alpha}$.

(e) *The function* $\alpha \mapsto \bar{\alpha}$ *is the only nonidentity ring homomorphism* $\mathbb{Q}[\sqrt{D}] \to \mathbb{Q}[\sqrt{D}]$.

Exercises

4.1.1.* Observe that $\mathbb{Q}[\sqrt{D}]$ is a two-dimensional vector space over $\mathbb{Q}$.

(a) Show that $\mathbb{Q}[\sqrt{D}] = \mathbb{Q} + \mathbb{Q}\alpha$ for any $\alpha \in \mathbb{Q}[\sqrt{D}] \setminus \mathbb{Q}$.

(b) Let $\beta \in \mathbb{Q}[\sqrt{D}]$. Show that $\gamma \mapsto \gamma\beta$ defines a $\mathbb{Q}$-linear transformation of $\mathbb{Q}[\sqrt{D}]$. Write down its matrix M_β with respect to your favorite basis of $\mathbb{Q}[\sqrt{D}]$, and compute the trace and determinant of M_β.

4.1.2. Show that $\mathbb{Q}[\sqrt{D}] = \mathbb{Q}[\sqrt{D'}]$ if and only if $D/D' = q^2$ for some $q \in \mathbb{Q}$.

4.1.3. Let $\mathbb{Q}[\sqrt{D}]$ be an imaginary quadratic field. Show that $\mathrm{N}\alpha > 0$ for all $\alpha \in \mathbb{Q}[\sqrt{D}] \setminus 0$.

4.1.4.* Let $\alpha \in \mathbb{Q}[\sqrt{D}]$. Prove that $\mathrm{N}\alpha = 1$ implies that there exists a $\beta \in \mathbb{Q}[\sqrt{D}]$ with $\alpha = \beta/\bar{\beta}$. Describe all such β.

4.1.5. Prove Prop. 4.1.3 (a)-(d).

4.1.6.* The proof of Prop. 4.1.3 (e) is more involved. Let $f : \mathbb{Q}[\sqrt{D}] \to \mathbb{Q}[\sqrt{D}]$ be a ring homomorphism different from the identity. Prove that $f(\alpha) = \bar{\alpha}$ by checking the following two claims:

(a) For each $q \in \mathbb{Q}$, $f(q) = q$; and

(b) $f(\sqrt{D}) = -\sqrt{D}$.

If you're familiar with Galois theory you will recognize this proof as a special case.

4.2 The Ring of Integers

We restate Def. 1.3.3 in the context of a general quadratic field.

4.2.1 Definition. *The **ring of integers** of $\mathbb{Q}[\sqrt{D}]$ is the set*

$$\mathcal{O} = \{\alpha \in \mathbb{Q}[\sqrt{D}] : \alpha^2 - t\alpha + n = 0 \text{ for some } t, n \in \mathbb{Z}\}$$
$$= \{\alpha \in \mathbb{Q}[\sqrt{D}] : \operatorname{Tr}\alpha, \operatorname{N}\alpha \in \mathbb{Z}\}.$$

The second equality is Prop. 4.1.3 (c). Though defined as a set, it is clear from the following explicit description that $\mathcal{O}$ is indeed a ring.

4.2.2 Theorem. *Assume that $D \in \mathbb{Z}$ is square-free. The set of integers in $\mathbb{Q}[\sqrt{D}]$ is a ring given by $\mathcal{O} = \mathbb{Z} + \mathbb{Z}\delta_0 = \mathbb{Z}[\delta_0]$, where*

$$\delta_0 = \begin{cases} \sqrt{D} & \text{for } D \equiv 2, 3 \pmod 4 \\ \frac{1+\sqrt{D}}{2} & \text{for } D \equiv 1 \pmod 4 \end{cases}$$

Proof. In either case δ_0 is in $\mathcal{O}$, since it satisfies a monic equation with coefficients in $\mathbb{Z}$, namely $x^2 - D = 0$ or $x^2 - x + (1 - D)/4 = 0$. The latter equation has coefficients in $\mathbb{Z}$ precisely when $D \equiv 1 \pmod 4$. It's easy to see that $\mathbb{Z} + \mathbb{Z}\delta_0 \subseteq \mathcal{O}$; we focus on proving the reverse inclusion.

Take an $\alpha = a + b\sqrt{D} \in \mathcal{O}$. By Def. 4.2.1, this means that $\operatorname{Tr}\alpha = 2a$ and $\operatorname{N}\alpha = a^2 - b^2D$ are both in $\mathbb{Z}$. Put $a = r/2, b = m/n$ for some $r, m, n \in \mathbb{Z}$ with $\gcd(m, n) = 1$. Then $4m^2D = n^2(r^2 - 4\operatorname{N}\alpha)$, so that $n^2 \mid 4m^2D$, and in fact $n^2 \mid 4D$ because $\gcd(m, n) = 1$. If p were an odd prime factor of n, we'd have $p^2 \mid D$, contradicting square-freeness of D. Thus, n has to be a power of 2. Since $4 \nmid D$, $n^2 \mid 4D$ implies $n^2 \mid 8$, hence $n = 1$ or 2. In either case, $b = s/2$ for some $s \in \mathbb{Z}$.

Since $a^2 - b^2D \in \mathbb{Z}$, we have $r^2 \equiv s^2D \pmod 4$. We consider two cases, bearing in mind that the only squares modulo 4 are 0 and 1:

(a) If $D \not\equiv 1 \pmod 4$, each pair (r, s) satisfies $r^2 \equiv s^2 \equiv 0 \pmod 4$. This implies that r and s are even integers, hence a and b are in $\mathbb{Z}$ and $\mathcal{O} \subseteq \mathbb{Z} + \mathbb{Z}\sqrt{D}$.

(b) If $D \equiv 1 \pmod 4$, then $r^2 \equiv s^2D \equiv s^2 \pmod 4$, which implies $r \equiv s \pmod 2$. Writing $r = s + 2k$ for $k \in \mathbb{Z}$ we see that

$$\alpha = a + b\sqrt{D} = \frac{r + s\sqrt{D}}{2} = \frac{s + 2k + s\sqrt{D}}{2} = k + s\frac{1 + \sqrt{D}}{2}.$$

Thus $\mathcal{O} \subseteq \mathbb{Z} + \mathbb{Z}\frac{1+\sqrt{D}}{2}$. ∎

By Prop. 4.1.3 (c), any $\alpha \in \mathcal{O} \setminus \mathbb{Z}$ is the root of a *unique* monic quadratic polynomial with coefficients in $\mathbb{Z}$. This suggests that an invariant of that polynomial is also an invariant of α.

4.2.3 Definition. *The **discriminant** of $\alpha \in \mathcal{O}$ is* $\operatorname{disc} \alpha = (\operatorname{Tr} \alpha)^2 - 4N\alpha$.

The proof of the following two claims is left as an exercise.

4.2.4 Proposition. *Let $\alpha \in \mathcal{O}$ and $\beta \in \mathbb{Z}[\alpha] = \mathbb{Z} + \mathbb{Z}\alpha \subseteq \mathcal{O}$. Then* $\operatorname{disc} \beta = c^2 \cdot \operatorname{disc} \alpha$ *for some* $c \in \mathbb{N}$.

4.2.5 Corollary. *Let $\alpha, \beta \in \mathcal{O}$.*

(a) If $\mathbb{Z}[\beta] \subseteq \mathbb{Z}[\alpha]$, then $\operatorname{disc} \alpha \mid \operatorname{disc} \beta$.
(b) If $\mathbb{Z}[\beta] = \mathbb{Z}[\alpha]$, then $\operatorname{disc} \alpha = \operatorname{disc} \beta$.

Part (b) says that $\operatorname{disc} \alpha$ depends only on the subring $R = \mathbb{Z}[\alpha] \subseteq \mathcal{O}$ that α generates, and so it makes sense to write $\operatorname{disc} R = \operatorname{disc} \alpha$. By part (a), larger subrings of $\mathcal{O}$ have smaller discriminants. In particular, $\operatorname{disc} \mathcal{O}$ should divide $\operatorname{disc} R$ for all subrings $R \subseteq \mathcal{O}$ of the form $\mathbb{Z}[\alpha]$.

4.2.6 Definition. *We put $D_F = \operatorname{disc} \mathcal{O} = \operatorname{disc} \delta_0$ and call it the **discriminant of the field** $F = \mathbb{Q}[\sqrt{D}]$.*

The following table summarizes the basic information about the ring of integers $\mathcal{O}$ of $F = \mathbb{Q}[\sqrt{D}]$, *with D square-free:*

$D \bmod 4$	δ_0, where $\mathcal{O} = \mathbb{Z}[\delta_0]$	Equation for δ_0	$D_F = \operatorname{disc} \mathcal{O}$
2, 3	$\sqrt{D}$	$\delta_0^2 - D = 0$	$4D$
1	$\frac{1+\sqrt{D}}{2}$	$\delta_0^2 - \delta_0 + \frac{1-D}{4} = 0$	D

Table 4.1 All you need to know about $\mathcal{O}$

A few remarks:

(a) The discriminant D_F of $\mathcal{O}$ is always square-free, except for a possible factor of 4. By Prop. 4.2.4, this is false for the discriminant of any strictly smaller subring $\mathbb{Z}[\alpha] \subsetneq \mathcal{O}$, which will be crucial to our proof of unique factorization of ideals in $\mathcal{O}$.
(b) We can always write $F = \mathbb{Q}[\sqrt{D_F}]$.
(c) If you don't like distinguishing between the two cases in Table 4.1, instead of δ_0 you can use the slightly more complicated

$$\delta_1 = \frac{D_F + \sqrt{D_F}}{2}, \text{ for which } \delta_1^2 - D_F\delta_1 + \frac{D_F^2 - D_F}{4} = 0.$$

Then $\mathcal{O} = \mathbb{Z}[\delta_1]$ regardless of the value of $D \bmod 4$.
(d) When proving general theorems about $\mathcal{O}$ we don't really care about the particular choice of δ with $\mathcal{O} = \mathbb{Z}[\delta]$; any δ with the right discriminant will do. Therefore, we fix the following notation:

NOTATIONAL CONVENTION

For the rest of the book, F denotes a quadratic field with discriminant D_F and ring of integers $\mathcal{O} = \mathbb{Z}[\delta]$. Such a δ must satisfy $\delta^2 - t\delta + n = 0$ for some $t, n \in \mathbb{Z}$ with $t^2 - 4n = D_F$.

Recall that we have already determined the units in the rings of integers of the following three fields:

$$F = \mathbb{Q}[i], \qquad D_F = -4, \quad \mathcal{O}^\times = \{1, i, -1, -i\}$$
$$F = \mathbb{Q}[\omega], \qquad D_F = -3, \quad \mathcal{O}^\times = \{1, \omega, \omega^2, \cdots, \omega^5\}$$
$$F = \mathbb{Q}[\sqrt{-5}], D_F = -20, \mathcal{O}^\times = \{1, -1\}$$

The first two are exceptional among imaginary quadratic fields.

4.2.7 Proposition. *If $D_F < -4$, then $\mathcal{O}^\times = \{\pm 1\}$.*

Proof. We know from Prop. 1.2.3 that $\varepsilon = a + b\delta \in \mathcal{O}^\times$ if and only if $N\varepsilon = \pm 1$. By Prop. 4.1.3 (a), this means that $a^2 + tab + nb^2 = \pm 1$.

If $b = 0$, then $a^2 = 1$ and $\varepsilon = a = \pm 1$. If $b \neq 0$, we divide through by b to get

$$\left(\frac{a}{b}\right)^2 + t\left(\frac{a}{b}\right) + n = \pm\frac{1}{b^2}.$$

If $D_F < -4$, this equation has no solution in $\mathbb{Z}$. Indeed, $-D_F/4 > 1 \geq \pm 1/b^2$, contradicting the fact that $-D_F/4$ is the minimum value of the function $x^2 + tx + n$ on $\mathbb{R}$. ∎

Exercises

4.2.1. Let $F = \mathbb{Q}[\sqrt{D}]$ for $D \in \{-47, -20, -13, 14, 57, 83\}$. Compute D_F and a δ with $\mathcal{O} = \mathbb{Z}[\delta]$, along with the corresponding t and n.

4.2.2. Let $F = \mathbb{Q}[\sqrt{D}]$, with D square-free. Determine, as a function of D (mod 4), the maximal power of 2 dividing D_F.

4.2.3. What is the smallest possible discriminant of a real quadratic field?

4.2.4. Find all $\delta' \in \mathcal{O}$ for which $\mathcal{O} = \mathbb{Z}[\delta']$.

4.2.5. Assume for this problem that $D < 0$, so that we can think of $\mathcal{O} = \mathbb{Z}[\delta]$ as a lattice in the complex plane. Compute the area of the fundamental parallelogram of this lattice (e.g., the one with vertices $0, 1, \delta, 1 + \delta$). Does this area depend on the choice of δ?

4.2.6. Prove Prop. 4.2.4 and Cor. 4.2.5.

4.2.7. Let $F = \mathbb{Q}[\sqrt{D}]$ with $D \equiv 1 \pmod 4$, and let $k \in \mathbb{Z}$. Show that the equation $N\alpha = k$ has a solution $\alpha \in \mathcal{O}$ if and only if the equation $x^2 - Dy^2 = 4k$ has a solution $x, y \in \mathbb{Z}$. When solving norm equations it thus suffices to look at equations of the form $x^2 - Dy^2 = a$.

4.2.8. If D_F has a positive prime divisor $p \equiv 3 \pmod 4$, then all units have norm 1.

4.2.9. Let $\alpha \in \mathcal{O}$. Prove that $|\mathcal{O}/\mathcal{O}\alpha| = |N\alpha|$.

4.2.10. Show that the ring of integers in an imaginary quadratic $\mathbb{Q}[\sqrt{D}]$ is a Euclid domain with the norm as its size function if and only if $D \in \{-1, -2, -3, -7, -11\}$.

Orders

This is the first in a series of blocks of exercises in which we develop the theory of subrings of $\mathcal{O}$. Their arithmetic is similar to that of $\mathcal{O}$, so long as we "stay away" from the conductor (defined below).

4.2.11. Let $R \subseteq \mathcal{O}$ be a subring.

(a) Prove that R is either $\mathbb{Z}$ or $\mathcal{O}_c = \mathbb{Z} + \mathbb{Z}c\delta$, where

$$c = |\mathcal{O}/R| = \sqrt{\frac{\operatorname{disc} R}{\operatorname{disc} \mathcal{O}}}.$$

(b) Prove that $\mathcal{O}_c = \mathbb{Z}[\delta_c]$ for any $\delta_c \in F$ satisfying $\delta_c^2 - t_c\delta_c + n_c = 0$ with $t_c^2 - 4n_c = c^2 D_F = \operatorname{disc} \mathcal{O}_c$. Determine t_c and n_c when $\delta_c = c\delta$.

(c) Show that $\mathcal{O}_c \subseteq \mathcal{O}_{c'}$ if and only if $c' \mid c$.

The subring $\mathcal{O}_c$ is called the **order** in F (or $\mathcal{O}$) of **conductor** c. That is why $\mathcal{O} = \mathcal{O}_1$ is sometimes called the **maximal order**.

4.2.12. Fill in the following table of invariants of orders $\mathcal{O}_c \subseteq \mathcal{O}$. In the fourth column, write any δ_c for which $\mathcal{O}_c = \mathbb{Z}[\delta_c]$, and in the fifth, the equation satisfied by δ_c.

D_F	c	$Disc\,\mathcal{O}_c$	δ_c	$\delta_c^2 - t_c\delta_c + n_c = 0$
-4	3	-36	$3i$	$\delta_c^2 + 9 = 0$
-35	5			
			$2 + 2\sqrt{13}$	
		112		
		-112		
				$\delta_c^2 - 4\delta_c + 11 = 0$

4.2.13. Let $\alpha, \beta \in F$ be linearly independent over $\mathbb{Q}$. The **ring of multipliers** of $\Lambda = \mathbb{Z}\alpha + \mathbb{Z}\beta$ is defined by

$$\mathcal{O}_\Lambda = \{\gamma \in F : \gamma\Lambda \subseteq \Lambda\}.$$

Define $j, k, l, m, c \in \mathbb{Z}$ by $\delta\alpha = (j/c)\alpha + (k/c)\beta, \delta\beta = (l/c)\alpha + (m/c)\beta$. These five integers are uniquely determined once we require $c > 0$ and $\gcd(j, k, l, m, c) = 1$. Prove that $\mathcal{O}_\Lambda = \mathcal{O}_c$.

4.2.14.* We define the ring of multipliers for an arbitrary lattice $\Lambda = \mathbb{Z}\alpha + \mathbb{Z}\beta \subset \mathbb{C}$ by $\mathcal{O}_\Lambda = \{\gamma \in \mathbb{C} : \gamma\Lambda \subseteq \Lambda\}$.

(a) Give an example of a Λ with $\mathcal{O}_\Lambda = \mathbb{Z}$.
(b) Assume for the rest of the exercise that $\mathcal{O}_\Lambda$ is strictly bigger than $\mathbb{Z}$. Show that $F = \mathbb{Q}[\beta/\alpha]$ is an imaginary quadratic field.
(c) Show that $\mathcal{O}_\Lambda$ is an order in F.
(d) Show that $\mathcal{O}_{\eta\Lambda} = \mathcal{O}_\Lambda$. Find an $\eta \in \mathbb{C} \setminus 0$ for which $\eta\Lambda \subset F$.

4.3 Unique Factorization of Ideals: The Road Map

The first big theorem of this book is Thm. 1.1.1: apart from 0 and $\mathcal{O}$ itself, each ideal of $\mathcal{O}$ can be written as a product of prime ideals, essentially uniquely. We will show this by adapting the proof of Unique Factorization in $\mathbb{Z}$ on p. 4. There, we deduced the existence of a factorization using the Well-Ordering Principle, and the uniqueness from Euclid's Lemma and cancellation. We need to generalize those three tools to ideals in $\mathcal{O}$.

Well-Ordering Principle: The principle doesn't extend verbatim to the set of ideals of $\mathcal{O}$; it's not even clear how to equip this set with a total ordering. A closer inspection, however, reveals that we've only used the following weaker, multiplicative version of the principle:

Let $\mathcal{S} \subseteq \mathbb{N}$ be a nonempty subset. There exists a "divisibility-minimal" element $k \in \mathcal{S}$, in the sense that if $l \mid k$ and $l \neq k$, then $l \notin \mathcal{S}$.

Unlike in the usual $\leq$-ordering, $\mathcal{S}$ can have several divisibility-minimal elements. Those in the set $\{d \mid n : d > 1\}$, for example, are precisely the prime factors of n. Replacing divisibility by inclusion $\supseteq$, we translate the multiplicative Well-Ordering Principle into a statement about ideals of $\mathbb{Z}$:

Let $\mathcal{S}$ be a nonempty set of ideals of $\mathbb{Z}$. There exists an element $\mathbb{Z}k \in \mathcal{S}$ for which $\mathbb{Z}l \supsetneq \mathbb{Z}k$ implies $\mathbb{Z}l \notin \mathcal{S}$. In other words, every nonempty set of ideals in $\mathbb{Z}$ has a largest element.

In Cor. 4.4.5 we will generalize this to ideals in $\mathcal{O}$.

Euclid's Lemma: You proved the analogue of Euclid's Lemma for ideals in Exer. 2.5.1. An ideal P of $\mathcal{O}$ is prime if and only if the following statement holds for any two ideals I, J of $\mathcal{O}$: $P \mid IJ$ implies $P \mid I$ or $P \mid J$.

Cancellation: Let I, J, K be ideals of $\mathcal{O}$ with $I \neq 0$. We'll need to cancel I in the identity $IJ = IK$, to deduce $J = K$. If I had an inverse for ideal multiplication, we could just multiply both sides by I^{-1}. Alas, only $\mathcal{O}$ has such an inverse (see Exer. 2.4.3 e), so we'll have to make the other inverses 'by hand'. We'll do this in Sec. 4.7 by suitably broadening the notion of ideal.

4.4 Noether Rings

We identify $\mathcal{O}$ with the standard lattice Λ_0 by means of the group isomorphism

(4.4.1) $$x + y\delta \leftrightarrow \begin{bmatrix} x \\ y \end{bmatrix}.$$

4.4.2 Proposition. *The isomorphism* (4.4.1) *identifies a nonzero ideal I of $\mathcal{O}$ with a sublattice of Λ_0.*

Proof. Suppose that there exists an ideal $I \neq 0$ of $\mathcal{O}$ that doesn't correspond to a sublattice (a subgroup of the form $\mathbb{Z}t_1 + \mathbb{Z}t_2$ with t_1 and t_2 linearly independent). By the classification of subgroups of Λ_0 in Thm. 3.1.6, I must correspond to a subgroup of the form $\mathbb{Z}t$. Let $\tau \in I$ correspond to the generator t. Then $\delta I \subseteq I$ implies that $\delta\tau \in \mathbb{Z}\tau$, i.e., $\delta\tau = n\tau$ for some $n \in \mathbb{Z}$. This is an equality in F, so we can cancel τ conclude that $\delta = n$, a contradiction. ∎

4.4.3 Proposition. *Let I be a nonzero ideal of $\mathcal{O}$.*

(a) $\mathcal{O}/I$ is finite.
(b) Any strictly ascending chain of ideals of $\mathcal{O}$ must be finite.

Proof. Both claims follow easily from our study of lattices.

(a) This is a special case of Prop. 3.4.5.
(b) A strictly ascending chain of ideals of $\mathcal{O}$ is a (finite or infinite) sequence of ideals satisfying $I_1 \subsetneq I_2 \subsetneq \cdots \subseteq \mathcal{O}$. We take the quotient of all ideals in this sequence by I_1 to get

$$0 = I_1/I_1 \subsetneq I_2/I_1 \subsetneq \cdots \subseteq \mathcal{O}/I_1.$$

This is a strictly ascending chain of subgroups of $\mathcal{O}/I_1$. By part (a), $\mathcal{O}/I_1$ is finite, so it has only finitely many subgroups. ∎

4.4.4 Corollary. *Every prime ideal of $\mathcal{O}$ is maximal.*

Proof. The Corollary follows from Prop. 4.4.3 (a) and the general criterion of Cor. 2.5.8. ∎

Let S be a collection of ideals of $\mathcal{O}$ partially ordered by inclusion $\subseteq$. We say that an ideal $I \in S$ is **largest in** S if it is not a proper subset of another $J \in S$. Such an ideal I would usually be called a maximal element of S; we introduce the new term to avoid confusing the statements "the ideal I is a maximal element in S" and "I is a maximal ideal of $\mathcal{O}$" in the sense of Def. 2.5.1. We now state a form of mathematical induction for ideals.

4.4.5 Corollary. *Any nonempty set S of ideals of $\mathcal{O}$ has a largest element.*

Proof. Assume the contrary. Take any ideal $I_1 \in \mathcal{S}$. By assumption, I_1 isn't a largest element in $\mathcal{S}$, so there exists an $I_2 \in \mathcal{S}$ with $I_1 \subsetneq I_2$. Repeating this reasoning yields an infinite strictly increasing chain $I_1 \subsetneq I_2 \subsetneq I_3 \subsetneq \cdots$, violating Prop. 4.4.3 (b). $\blacksquare$

4.4.6 Corollary. *Every ideal $I \subsetneq \mathcal{O}$ is contained in a maximal ideal of $\mathcal{O}$.*

Proof. This follows from Cor. 4.4.5 applied to the nonempty set $\{J : J$ is an ideal of $\mathcal{O}$ and $I \subseteq J \subsetneq \mathcal{O}\}$. $\blacksquare$

Using Zorn's Lemma, the statement of the corollary can be proved in any ring. Condition (b) of Prop. 4.4.3 captures the idea that $\mathcal{O}$ isn't 'too big', and merits being singled out for a definition. It is named after Emmy Noether, the mother of the modern theory of rings and ideals.

4.4.7 Definition. *A ring R is said to be* **Noether**[1] *if it satisfies the* **Ascending Chain Condition***: every strictly ascending chain of ideals $I_1 \subsetneq I_2 \subsetneq \cdots$ is finite.*

It's good, for contrast, to see a ring that isn't Noether.

4.4.8 Example. Consider the polynomial ring in countably many variables $R = \mathbb{C}[x_1, x_2, x_3, \ldots]$. Let's be clear on what this means: we have infinitely many variables at our disposal, but any one element of R is an ordinary polynomial with coefficients in $\mathbb{C}$ that involves only *finitely* many of the variables, e.g., $-ix_1^2 x_3 + 7x_{18}^5 - \pi x_{1563}$. The ring structure is given by the usual addition and multiplication of polynomials. Then

$$\langle x_1 \rangle \subsetneq \langle x_1, x_2 \rangle \subsetneq \langle x_1, x_2, x_3 \rangle \subsetneq \cdots$$

is an infinite strictly ascending chain of ideals. $\square$

Exercises

4.4.1. Verify that the following statement is equivalent to the Ascending Chain Condition: every (not necessarily strictly) ascending chain of ideals $I_1 \subseteq I_2 \subseteq \cdots$ eventually stabilizes, i.e., there exists a k_0 such that $I_k = I_{k_0}$ for all $k \geq k_0$.

4.4.2. Find all strictly ascending chains of ideals in $\mathbb{Z}[i]$ that start with $\mathbb{Z}[i] \cdot (45 - 60i)$.

4.4.3. Show that $\mathbb{Z}$ is a Noether ring.

4.4.4.* Let R be a ring. Prove that a ring R is Noether if and only if every ideal I of R is finitely generated: there exist $x_1, \ldots, x_n \in I$ such that $I = Rx_1 + \cdots + Rx_n$.

---

[1] More commonly, Noetherian.

4.4.5. Prove that each element in a Noether integral domain can be written as a product of irreducible elements (as we know, not necessarily uniquely).

4.4.6.* Let R be a Noether ring and $\varphi \colon R \to R$ a surjective ring homomorphism. Prove that φ has to be injective.

4.5 Standard Form of an Ideal

Given a sublattice $\Lambda \subseteq \Lambda_0$, we used column reduction to write $\Lambda = \gamma \Lambda_0$ for a particularly simple matrix γ. After identifying $\mathcal{O}$ with Λ_0 via (4.4.1), this procedure allows us to explicitly describe all ideals of $\mathcal{O}$.

4.5.1 Proposition. *For any ideal $I \subseteq \mathcal{O}$, there exist $a, b, d \in \mathbb{Z}$ such that the following are true:*

(a) $I = d(\mathbb{Z}a + \mathbb{Z}(-b + \delta))$, *and*
(b) $b^2 - tb + n \equiv 0 \pmod{a}$, *or, equivalently,* $a \mid \mathrm{N}(-b + \delta)$

Conversely, any $I \subseteq \mathcal{O}$ satisfying (a) and (b) is an ideal.

The expression in (a) is said to be a **standard form** of the ideal I. Exer. 4.5.2 shows that the standard form becomes unique once we require that $d > 0$ and $0 \le b < a$.

Proof. By Prop. 4.4.2, the isomorphism (4.4.1) identifies the ideal I with a sublattice $\Lambda = \gamma \Lambda_0$ for some $\gamma \in M_{2 \times 2}(\mathbb{Z})$. Column-reducing γ produces $a', b', d \in \mathbb{Z}$ for which

$$\Lambda = \begin{bmatrix} a' & b' \\ 0 & d \end{bmatrix} \Lambda_0 = \mathbb{Z}\begin{bmatrix} a' \\ 0 \end{bmatrix} + \mathbb{Z}\begin{bmatrix} b' \\ d \end{bmatrix}, \text{ and correspondingly}$$
$$I = \mathbb{Z}a' + \mathbb{Z}(b' + d\delta).$$

Now we bring into play the fact that I absorbs multiplication by δ. We apply this to the basis $\{a', b' + d\delta\}$ of I to prove the two conditions of the proposition.

(a) Since I is an ideal, $\delta a' \in I$, hence there exist $h, k \in \mathbb{Z}$ such that

$$\delta a' = ha' + k(b' + d\delta) = (ha' + kb') + kd\delta.$$

We conclude that $a' = kd$ and $ha' + kb' = 0$. Putting $a = k, b = h$ gives $b' = -bd$, and, as desired,

$$I = \mathbb{Z}da + \mathbb{Z}(-bd + d\delta) = d(\mathbb{Z}a + \mathbb{Z}(-b + \delta)).$$

(b) It is sufficient to prove part (b) in the case $d = 1$. Since $\delta I \subseteq I$, we have $\delta(-b + \delta) = -b\delta + \delta^2 = -n + (t - b)\delta \in I$, where the last equality follows from $\delta^2 - t\delta + n = 0$. Thus, there exist $j, l \in \mathbb{Z}$ with

$$-n + (t - b)\delta = ja + l(-b + \delta) = (ja - lb) + l\delta.$$

Comparing rational and irrational parts, we get $-n = ja - lb$ and $t - b = l$. Multiplying the second equality by $-b$ gives $b^2 - tb = -lb = -ja - n$. This proves the desired congruence,

$$b^2 - tb + n = -ja \equiv 0 \pmod{a}.$$

By Prop. 4.1.3 (a), the left side is just $N(-b + \delta)$.

For the converse, we reverse these calculations to show that $\delta I \subseteq I$. ∎

4.5.2 Example. Let's illustrate the argument of Prop. 4.5.1 in the concrete case $\mathcal{O} = \mathbb{Z}[i]$ and $I = \mathcal{O}(2 + i) = (\mathbb{Z} + \mathbb{Z}i)(2 + i) = \mathbb{Z}(2 + i) + \mathbb{Z}(-1 + 2i)$. To put I into standard form, we column-reduce:

$$\begin{bmatrix} 2 & -1 \\ 1 & 2 \end{bmatrix} \xrightarrow{C_2 - 2C_1} \begin{bmatrix} 2 & -5 \\ 1 & 0 \end{bmatrix} \xrightarrow[-C_1]{C_1 \circlearrowright C_2} \begin{bmatrix} 5 & 2 \\ 0 & 1 \end{bmatrix}.$$

One standard form of I is thus $\mathbb{Z} \cdot 5 + \mathbb{Z}(2 + i)$. Fig. 4.1 shows two fundamental parallelograms of the lattice $I \subset \mathbb{C}$, one corresponding to the initial basis $\{2 + i, -1 + 2i\}$ of I, the other to the basis $\{5, 2 + i\}$ of the standard form. Because of the zero in the first column of the column-reduced form, the second parallelogram has its base along the x-axis.

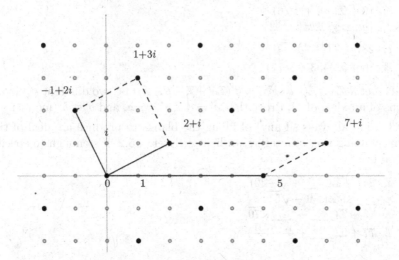

Fig. 4.1 Standard form of the ideal $\mathbb{Z}[i](2+i) \subseteq \mathbb{Z}[i]$. All points are elements of $\mathbb{Z}[i]$, while the bold ones are in I

Both fundamental parallelograms contain five points in $\mathbb{Z}[i]$, not counting the dashed lines or their endpoints. Exer. 3.1.6 implies that $|\mathcal{O}/I| = 5$, which is also the norm of the generator $2 + i$. This is not an accident, as we saw in Exer. 4.2.9. □

4.5.3 Example. When is 19 contained in an ideal $I = d(\mathbb{Z}a + \mathbb{Z}(-b + \sqrt{-37}))$ of the ring $\mathbb{Z}[\sqrt{-37}]$? That happens when there exist $k, l \in \mathbb{Z}$ such that $19 = d(ka + l(-b + \sqrt{-37}))$. Comparing coefficients, we see that $19 = dka$.

If $k = 19$, then $d = a = 1$ and $I = \mathbb{Z}[\sqrt{-37}]$.

If $d = 19$, then $a = 1$ and $I = \mathbb{Z} \cdot 19 + \mathbb{Z} \cdot 19\sqrt{-37}$, which is the principal ideal $\mathbb{Z}[\sqrt{-37}] \cdot 19$.

Finally, if $a = 19$, we look for ideals of the form $\mathbb{Z} \cdot 19 + \mathbb{Z}(-b + \sqrt{37})$ with $b^2 + 37 \equiv 0 \pmod{19}$, or $b^2 \equiv 1 \pmod{19}$. Taking $b = \pm 1$, we get the possible ideals:

$$P_1 = \mathbb{Z} \cdot 19 + \mathbb{Z}(1 + \sqrt{-37})$$
$$P_2 = \mathbb{Z} \cdot 19 + \mathbb{Z}(-1 + \sqrt{-37}) = \mathbb{Z} \cdot 19 + \mathbb{Z}(1 - \sqrt{-37}) = \bar{P}_1. \qquad \square$$

Exercises

4.5.1. Use Prop. 4.5.1 to decide which of the following are ideals in the corresponding ring of integers:

(a) $\mathbb{Z} \cdot 429 + \mathbb{Z}(87 + 3\sqrt{-17})$
(b) $\mathbb{Z} \cdot 10 + \mathbb{Z}(-2 + 4\sqrt{6})$
(c) $\mathbb{Z} \cdot 120 + \mathbb{Z}(\frac{105 + 5\sqrt{-39}}{2})$
(d) $\mathbb{Z} \cdot 287 + \mathbb{Z}(\frac{119 + \sqrt{37}}{2})$
(e) $\mathbb{Z} \cdot 86 + \mathbb{Z}(-53 + \sqrt{17})$

4.5.2. Let $d(\mathbb{Z}a + \mathbb{Z}(-b + \delta)) = d'(\mathbb{Z}a' + \mathbb{Z}(-b' + \delta))$ be two different standard forms of an ideal of $\mathcal{O}$. Prove that $d' = \pm d, a' = \pm a$, and $b' \equiv b \pmod{a}$.

4.5.3. Find all ways (if any) of filling the blanks to produce an ideal of the corresponding ring of integers. In light of Exer. 4.5.2, it's enough to specify b mod a:

(a) $\mathbb{Z} \cdot 247 + \mathbb{Z}(\underline{} + \sqrt{-22})$
(b) $\mathbb{Z} \cdot \underline{} + \mathbb{Z}(-9 + \sqrt{51})$
(c) $\mathbb{Z} \cdot 13 + \mathbb{Z}(\underline{} + \underline{}\sqrt{70})$
(d) $\mathbb{Z} \cdot 77 + \mathbb{Z}(\underline{} + \frac{1 + \sqrt{-31}}{2})$

Orders

Let $\mathcal{O}_c \subseteq \mathcal{O}$ be the order of conductor c.

4.5.4. Prove the analog of Prop. 4.5.1 for orders: I is an ideal of $\mathcal{O}_c$ if and only if we can find $a, b, d \in \mathbb{Z}$ satisfying

$$I = d(\mathbb{Z}a + \mathbb{Z}(-b + c\delta)) \text{ with } b^2 - (ct)b + c^2 n \equiv 0 \pmod{a}.$$

4.5.5. Consider the principal ideal $c\mathcal{O}$ of $\mathcal{O}$. We write the generator on the left to avoid confusion with $\mathcal{O}_c$.

(a) Show that $c\mathcal{O} \subseteq \mathcal{O}_c$.
(b) Show that $c\mathcal{O}$ contains any other ideal of $\mathcal{O}_c$ that is also an ideal of $\mathcal{O}$, i.e., absorbs multiplication by the bigger ring $\mathcal{O}$.
(c) Is $c\mathcal{O}$ a principal ideal of $\mathcal{O}_c$?

4.5.6. Let $I = d(\mathbb{Z}a + \mathbb{Z}(-b + c\delta))$ be an ideal of $\mathcal{O}_c$. We know from Exer. 4.2.13 that the ring of multipliers $\mathcal{O}_I = \{\gamma \in F : \gamma I \subseteq I\}$ is an order $\mathcal{O}_{c'}$. Show that $c' = c/\gcd(a, c)$.

4.6 The Ideal Norm

We will extend the notion of the norm from elements of F to ideals of $\mathcal{O}$. By Exer. 4.2.9, $|N\alpha| = |\mathcal{O}/\mathcal{O}\alpha|$ for any $\alpha \in \mathcal{O}$. By Prop. 4.4.3, $\mathcal{O}/I$ is finite for any ideal $I \neq 0$. Those two facts motivate the definition we're after.

4.6.1 Definition. *The* **norm** *of an ideal* $I \neq 0$ *of* $\mathcal{O}$ *is the natural number* $NI = |\mathcal{O}/I|$.

4.6.2 Example. Proposition 3.4.5 shows how to compute the size of the quotient of two lattices. We apply it to find the norm of ideal P_1 from Ex. 4.5.3:

$$NP_1 = |\mathbb{Z}[\sqrt{-37}]/(\mathbb{Z} \cdot 19 + \mathbb{Z}(-1 + \sqrt{-37}))|$$
$$= \left|\Lambda_0 / \begin{bmatrix} 19 & -1 \\ 0 & 1 \end{bmatrix} \Lambda_0\right| = \left|\det \begin{bmatrix} 19 & -1 \\ 0 & 1 \end{bmatrix}\right| = 19. \qquad \square$$

The ideal norm is essential for reducing multiplicative questions about ideals of $\mathcal{O}$ to analogous but easier questions about natural numbers.

4.6.3 Proposition. *Let I and J be nonzero ideals of $\mathcal{O}$. If $I \mid J$, then $NI \mid NJ$.*

Proof. $I \mid J$ just means that $I \supseteq J$, so $\alpha + J \mapsto \alpha + I$ is a well-defined, surjective ring homomorphism $\varphi : \mathcal{O}/J \to \mathcal{O}/I$. By the First Isomorphism Theorem, $\mathcal{O}/I \cong (\mathcal{O}/J)/\ker\varphi$. Counting elements, we get $|\mathcal{O}/I| \cdot |\ker\varphi| = |\mathcal{O}/J|$, so that $NI = |\mathcal{O}/I|$ divides $NJ = |\mathcal{O}/J|$. ∎

4.6.4 Example. Let's show that the ideals

$$I = \mathbb{Z} \cdot 11 + \mathbb{Z}(3 + \sqrt{31}), J = \mathbb{Z} \cdot 6 + \mathbb{Z}(1 + \sqrt{31})$$

of $\mathbb{Z}[\sqrt{31}]$ are relatively prime. Since $I + J$ divides both I and J, Prop. 4.6.3 shows that $N(I + J) \mid \gcd(NI, NJ) = \gcd(11, 6) = 1$, so that indeed $I + J = \mathbb{Z}[\sqrt{31}]$. $\square$

Next, we generalize to the ideal norm the formula $N\alpha = \alpha\bar\alpha$, which holds for elements $\alpha \in F$. This is key to the last remaining point in the program of Sec. 4.3 for proving the unique factorization of ideals, namely the cancellation for ideal multiplication.

4.6.5 Theorem. *Let I be a nonzero ideal of $\mathcal{O}$. Then $I\bar I = \mathcal{O} \cdot NI$, the principal ideal generated by NI viewed as an element of $\mathcal{O}$.*

Proof. By Prop. 4.5.1, for any such I we can find $a,b,d,j \in \mathbb{Z}, a,d \neq 0$, such that

$$I = d(\mathbb{Z}a + \mathbb{Z}(-b+\delta)) \text{ with } b^2 - tb + n = ja.$$

Putting $I' = \mathbb{Z}a + \mathbb{Z}(-b+\delta)$, we have that

$$I\bar I = dI'(d\bar I') = d^2(I'\bar I') = (N\bar d)(I'\bar I').$$

It is thus enough to prove the Theorem for I', so we assume that $d = 1$.

From Def. 4.1.2 we have $\bar\delta = t - \delta$, hence $I\bar I = (\mathbb{Z}a + \mathbb{Z}(-b+\delta))(\mathbb{Z}a + \mathbb{Z}(-b+t-\delta))$. We multiply the two ideals as in Ex. 3.3.1, and get that $I\bar I$ is the $\mathbb{Z}$-linear span of the set

$$\{a^2, \quad -ab + a\delta, \quad -ab + at - a\delta, \quad b^2 - tb + n\}.$$

To find a basis for $I\bar I$ we perform a column reduction, starting with

(4.6.6)
$$\begin{bmatrix} a^2 & -ab & -ab+at & ja \\ 0 & a & -a & 0 \end{bmatrix} \xrightarrow[C_2 \circlearrowleft C_4]{-(C_3+C_2)} \begin{bmatrix} a^2 & ja & 2ab-at & -ab \\ 0 & 0 & 0 & a \end{bmatrix}$$

The Column Reduction Algorithm instructs us to reduce the 1×3 submatrix $\begin{bmatrix} a^2 & ja & 2ab-at \end{bmatrix}$ next. Exercise 3.2.5 guarantees that the result is $\begin{bmatrix} 0 & 0 & \gcd(a^2,ja,2ab-at) \end{bmatrix}$, even though, in the absence of concrete numbers, we can't specify the column operations that get us there.

We claim that $\gcd(a^2, ja, 2ab - at) = a$, or, equivalently, that $\gcd(a, j, 2b - t) = 1$. If $p \in \mathbb{Z}$ were a prime dividing a, j and $2b - t$, then

(4.6.7)
$$p^2 \mid (2b-t)^2 - 4ja = 4b^2 - 4bt + t^2 - 4b^2 + 4bt - 4n$$
$$= t^2 - 4n = D_F.$$

By remark (a) after Table 4.1, D_F is not divisible by squares other than possibly 4, so p must be 2. You are invited to dismiss this possibility, and to conclude that $\gcd(a^2, ja, 2ab - t) = a$. This justifies the first step in the following continuation of column reduction (4.6.6):

$$\begin{bmatrix} a^2 & ja & 2ab-at & -ab \\ 0 & 0 & 0 & a \end{bmatrix} \longrightarrow \begin{bmatrix} 0 & 0 & a & -ab \\ 0 & 0 & 0 & a \end{bmatrix} \longrightarrow \begin{bmatrix} 0 & 0 & a & 0 \\ 0 & 0 & 0 & a \end{bmatrix}.$$

We conclude that $I\bar{I} = \mathbb{Z}a + \mathbb{Z}a\delta = \mathcal{O}a = \mathcal{O} \cdot NI$, since $NI = |\mathcal{O}/I| = \left|\det \begin{bmatrix} a & -b \\ 0 & 1 \end{bmatrix}\right| = a$. ∎

The proof of $I\bar{I} = \mathcal{O} \cdot NI$ pivots on the fact that $D_F = \operatorname{disc} \mathcal{O}$ is essentially square-free. This is false for the discriminant of any strictly smaller subring $\mathbb{Z}[\alpha] \subsetneq \mathcal{O}$, by Prop. 4.2.4. That is why the full ring $\mathcal{O}$ is the right place to look for unique prime factorization of ideals.

4.6.8 Corollary. *For any nonzero ideals I and J of $\mathcal{O}$, $N(IJ) = NI \cdot NJ$.*

Proof. This is immediate from Thm. 4.6.5. When $I + J = \mathcal{O}$, the Corollary also follows from the Chinese Remainder Theorem for rings, Prop. 2.4.8. ∎

We finally have the tools to prove cancellation for ideal multiplication.

4.6.9 Corollary. *Let I, J and K be ideals of $\mathcal{O}$, with $I \neq 0$. If $IJ = IK$, then $J = K$.*

Proof. The corollary follows from the following chain of implications:

$$IJ = IK \;\Rightarrow\; (I\bar{I})J = (I\bar{I})K \;\Rightarrow\; (NI)J = (NI)K \;\Rightarrow\; J = K.$$

The last implication follows by scaling both sides by NI^{-1}. ∎

See Exer. 4.6.9 for an example where cancellation fails in a subring of $\mathcal{O}$.

Exercises

4.6.1. Finish the proof of Thm. 4.6.5 by showing that the divisibility (4.6.7) is impossible when $p = 2$.

4.6.2. If $I \subseteq J$ are ideals of $\mathcal{O}$ and $NI = NJ$, then $I = J$.

4.6.3. Show that $NI = |d^2 a|$ for an ideal $I = d(\mathbb{Z}a + \mathbb{Z}(-b + \delta))$ of $\mathcal{O}$.

4.6.4. Let P be a maximal (i.e., nonzero prime) ideal of $\mathcal{O}$. Prove that $NP = p$ or p^2 for the unique prime $p \in \mathbb{N} \cap P$.

4.6.5. Let I be an ideal of $\mathcal{O}$. Show that $NI = \gcd(N\alpha : \alpha \in I)$.

4.6.6.* Let $I_k = d_k(\mathbb{Z}a_k + \mathbb{Z}(-b_k + \delta)), k = 1, 2$, be two ideals of $\mathcal{O}$ satisfying $\gcd(a_1, a_2) = 1$. Use the Chinese Remainder Theorem to choose a $b \in \mathbb{Z}$ satisfying $b \equiv b_1 \pmod{a_1}, b \equiv b_2 \pmod{a_2}$. Prove the following formula for multiplying such ideals:

$$I_1 I_2 = d_1 d_2(\mathbb{Z}(a_1 a_2) + \mathbb{Z}(-b + \delta)).$$

4.6.7.* Let I and J be ideals of $\mathcal{O}$ with $I \mid J$. Put $K = \{\alpha \in \mathcal{O} : \alpha I \subseteq J\}$. Show that K is an ideal of $\mathcal{O}$, and that $J = IK$.

Orders

Let $\mathcal{O}_c \subseteq \mathcal{O}$ be the order of conductor c.

4.6.8. Let $I = d(\mathbb{Z}a + \mathbb{Z}(-b+c\delta))$ be an ideal of $\mathcal{O}_c$ in standard form. We define the **norm of I relative to $\mathcal{O}_c$** by $\mathrm{N}_c I = |\mathcal{O}_c/I|$.

(a) Give a formula for $\mathrm{N}_c I$ in terms of a, b, c and d.
(b) For $\alpha \in \mathcal{O}_c$, show that $\mathrm{N}_c(\mathcal{O}_c \alpha) = \mathrm{N}(\mathcal{O}\alpha) = |\mathrm{N}\alpha|$.

4.6.9. Consider the field $\mathbb{Q}[\sqrt{-7}]$ and its order $\mathcal{O}_2 = \mathbb{Z}[\sqrt{-7}]$.

(a) Show that $I = \mathbb{Z} \cdot 2 + \mathbb{Z}(1 + \sqrt{-7})$ is an ideal of $\mathcal{O}_2$.
(b) Show that $I \cdot \bar{I} \neq \mathcal{O}_2 \cdot |\mathcal{O}_2/I|$.
(c) Show that I can't always be cancelled, by finding ideals J and K of $\mathcal{O}_2$ for which $IJ = IK$ but $J \neq K$.

4.6.10. Let $I = d(\mathbb{Z}a + \mathbb{Z}(-b+c\delta))$ be an ideal of $\mathcal{O}_c$. Show that the following are equivalent:

(a) $\gcd(a, c) = 1$.
(b) I can always be cancelled.
(c) $\mathcal{O}_I = \mathcal{O}_c$, where $\mathcal{O}_I$ is the ring of multipliers of I (see Exer. 4.5.6). In other words, I doesn't absorb multiplication by any order bigger than $\mathcal{O}_c$.

An ideal of $\mathcal{O}_c$ satisfying these conditions is called **invertible**.

4.6.11. Let I be an invertible ideal of $\mathcal{O}_c$. Prove that $I \cdot \bar{I} = \mathcal{O}_c \cdot \mathrm{N}_c I$, and that $\mathrm{N}_c I = \gcd(\mathrm{N}\alpha : \alpha \in I)$.

4.6.12. Let I be an ideal of $\mathcal{O}_c$ and J an ideal of $\mathcal{O}$ with $\gcd(\mathrm{N}_c I, c) = \gcd(\mathrm{N}J, c) = 1$.

(a) Compute the sizes of the lattice quotients $\mathcal{O}/\mathcal{O}_c$ and $\mathcal{O}_c/c\mathcal{O}$.
(b) Use Exer. 3.4.6 to show that $J + \mathcal{O}_c = \mathcal{O}$ and $I + c\mathcal{O} = \mathcal{O}_c$. Conclude from the latter that $Ic\mathcal{O} = I \cap c\mathcal{O}$.
(c) Deduce the isomorphisms of quotient rings $\mathcal{O}/J \cong \mathcal{O}_c/(J \cap \mathcal{O}_c)$ and $\mathcal{O}_c/I \cong \mathcal{O}/I\mathcal{O}$.

4.6.13. Consider the sets of ideals,

$$\mathbb{I}_c(c)^+ = \{I \text{ an ideal of } \mathcal{O}_c : \gcd(\mathrm{N}_c I, c) = 1\},$$
$$\mathbb{I}_1(c)^+ = \{J \text{ an ideal of } \mathcal{O} : \gcd(\mathrm{N}J, c) = 1\}.$$

The subscript is the conductor of the order in which the ideals lie; the number in parentheses appears in the norm condition.

Deduce from Exer. 4.6.12 that the assignments $I \mapsto I\mathcal{O}$ and $J \mapsto J \cap \mathcal{O}_c$ define mutually inverse bijections between $\mathbb{I}_c(c)^+$ and $\mathbb{I}_1^+(c)$. Show that these bijections preserve norms and ideal multiplication.

4.7 Fractional Ideals

Ideal multiplication is an associative and commutative operation on the set $\mathbb{I}_F^+$ of nonzero ideals of $\mathcal{O}$. The only invertible element in $\mathbb{I}_F^+$ is its identity $\mathcal{O}$ (see Exer. 2.4.3 e). Despite being so poor in inverses, $\mathbb{I}_F^+$ satisfies cancellation. To resolve this discrepancy, we enlarge $\mathbb{I}_F^+$ by the inverses of all nonzero ideals to a construct a group $\mathbb{I}_F$. From Thm. 4.6.5 we know that $I\bar{I} = \mathcal{O} \cdot NI$, so it's tempting to put $I^{-1} = (1/NI) \cdot \bar{I}$.

4.7.1 Example. Consider the ideal $I = \mathbb{Z}[i] \cdot 2$ of $\mathbb{Z}[i]$. With the above (provisional) definition, $I^{-1} = (1/2)\mathbb{Z}[i] = \mathbb{Z} \cdot 1/2 + \mathbb{Z} \cdot i/2$, which is a lattice, but is not contained inside $\mathbb{Z}[i]$. $\qquad\qquad\qquad\qquad\qquad\qquad\qquad\square$

A ideal I of $\mathcal{O}$ is a subgroup which absorbs multiplication by $\mathcal{O}$. We arrive at the notion of a fractional ideal by dropping the requirement that $I \subseteq \mathcal{O}$.

4.7.2 Definition. *A **fractional ideal** in F is an additive subgroup $\mathscr{I}$ of F that satisfies both of the following conditions:*

(a) $\mathscr{I} = \mathbb{Z}\alpha + \mathbb{Z}\beta$, where $\alpha, \beta \in F$ are linearly independent over $\mathbb{Z}$;
(b) $\delta\mathscr{I} \subseteq \mathscr{I}$.

Cursive letters will generally stand for fractional ideals. The set of all fractional ideals is denoted $\mathbb{I}_F$. A few remarks:

(a) Fractional ideals are a genuinely new class of objects. They are *not* ideals of $\mathcal{O}$ with some extra property called "fractionality"; indeed, they need not be subsets of $\mathcal{O}$ at all. Nor are fractional ideals simply ideals of the field F (of which there are exactly two).
(b) The field F itself is *not* a fractional ideal, since it's not of the form $\mathbb{Z}\alpha + \mathbb{Z}\beta$.
(c) A fractional ideal contained in $\mathcal{O}$ is just an ideal of $\mathcal{O}$ in the usual sense; in other words, $\mathbb{I}_F^+ \subseteq \mathbb{I}_F$. We will sometimes, for emphasis, call ideals of $\mathcal{O}$ **integral ideals**, and denote them, as before, with upright rather than cursive letters.

4.7.3 Example. For any $\alpha \in F \setminus 0$, the set

$$\mathcal{O}\alpha = \{\beta\alpha : \beta \in \mathcal{O}\} = \mathbb{Z}\alpha + \mathbb{Z}\delta\alpha$$

is a fractional ideal. Fractional ideals of this form are called **principal fractional ideals**. $\qquad\qquad\qquad\qquad\qquad\qquad\qquad\qquad\qquad\qquad\qquad\square$

4.7.4 Proposition. *An $\mathscr{I} \subseteq F$ is a fractional ideal if and only if there exists an $e \in \mathbb{Z}$ such that $e\mathscr{I}$ is an ideal of $\mathcal{O}$.*

Proof. Let $\mathscr{I} = \mathbb{Z}\alpha + \mathbb{Z}\beta$. Taking the common denominator, we can find $a, b, c, d, e \in \mathbb{Z}$ for which $\alpha = (a + b\sqrt{D})/e, \beta = (c + d\sqrt{D})/e$. Multiplying by e clears denominators, so $e\mathscr{I} \subseteq \mathcal{O}$. More than a mere subgroup, $e\mathscr{I}$ is an ideal of $\mathcal{O}$, since $\delta(e\mathscr{I}) = e(\delta\mathscr{I}) \subseteq e\mathscr{I}$. The converse is trivial. $\qquad\blacksquare$

We will refer to any e from the proposition as an (integer) **denominator** of $\mathscr{I}$. Combining Prop. 4.7.4 and Prop. 4.5.1 yields the following statement.

4.7.5 Corollary. *Fractional ideals are precisely the subsets of F of the form $q(\mathbb{Z}a + \mathbb{Z}(-b + \delta))$ for some $q \in \mathbb{Q}^{\times}$ and $a, b \in \mathbb{Z}$ satisfying $a \neq 0$ and $b^2 - tb + n \equiv 0 \pmod{a}$.*

Now that we have an explicit description of its elements, we proceed to define a group structure on $\mathbb{I}_F$.

4.7.6 Proposition. *Let $\mathscr{I}, \mathscr{J} \in \mathbb{I}_F$ and let $k, l \in \mathbb{Z}$ be such that $k\mathscr{I}, l\mathscr{J} \subseteq \mathcal{O}$. The formula*

$$\mathscr{I} \cdot \mathscr{J} = \frac{1}{kl}(k\mathscr{I}) \cdot (l\mathscr{J})$$

gives a well-defined operation on $\mathbb{I}_F$ extending the usual ideal multiplication on $\mathbb{I}_F^+$. Under this operation, $\mathbb{I}_F$ is an abelian group. Moreover, ideal norm extends to a homomorphism of multiplicative groups $\mathbb{I}_F \to \mathbb{Q}_{>0}^{\times}$ by putting $\mathrm{N}(\frac{1}{e}I) = \frac{1}{e^2}\mathrm{N}I$ for $n \in \mathbb{N}, I \in \mathbb{I}_F^+$.

Proof. The definition of $\mathscr{I} \cdot \mathscr{J}$ isn't circular, since the product on the right is the usual product of integral ideals $k\mathscr{I}$ and $l\mathscr{J}$. Let $k', l' \in \mathbb{Z}$ be another pair of denominators for $\mathscr{I}$ and $\mathscr{J}$, respectively. Then $kk'\mathscr{I} \subseteq \mathcal{O}$ and $ll'\mathscr{J} \subseteq \mathcal{O}$ as well. The associativity and commutativity of the usual ideal multiplication give

$$(k'l')(k\mathscr{I})(l\mathscr{J}) = (kk'\mathscr{I})(ll'\mathscr{J}) = (kl)(k'\mathscr{I})(l'\mathscr{J}).$$

Scaling by $1/kk'll'$, we get

$$\frac{1}{kl}(k\mathscr{I}) \cdot (l\mathscr{J}) = \frac{1}{k'l'}(k'\mathscr{I}) \cdot (l'\mathscr{J}),$$

which shows that the product of fractional ideals doesn't depend on the choice of denominators.

This operation obviously extends the multiplication on $\mathbb{I}_F^+$, and is associative and commutative with identity $\mathcal{O}$. It remains to establish the existence of inverses. Take $\mathscr{I} \in \mathbb{I}_F$ and $e \in \mathbb{Z}$ with $e\mathscr{I} \subseteq \mathcal{O}$. By Thm. 4.6.5, $(e\mathscr{I}) \cdot (e\bar{\mathscr{I}}) = \mathcal{O} \cdot \mathrm{N}(e\mathscr{I})$, so $\mathscr{I} \cdot (e^2/\mathrm{N}(e\mathscr{I}) \cdot \bar{\mathscr{I}}) = \mathcal{O}$, which shows that

$$\mathscr{I}^{-1} = \frac{e^2}{\mathrm{N}(e\mathscr{I})} \cdot \bar{\mathscr{I}}.$$

We leave the claim about the norm as an exercise. ∎

The construction of $\mathbb{I}_F$ from $\mathbb{I}_F^+$ is analogous to enlarging $\mathbb{N}$ by the inverses of all its elements to construct the positive rationals.

Exercises

4.7.1. Let $\mathbb{I}_\mathbb{Q}^+$ be the set of all nonzero ideals of $\mathbb{Z}$ equipped with ideal multiplication. Imitate the construction of this section to make a group $\mathbb{I}_\mathbb{Q} \supset \mathbb{I}_\mathbb{Q}^+$. Let $f : \mathbb{I}_\mathbb{Q}^+ \to G$ be a multiplication-preserving function to a group G. Show that f has a unique extension to a group homomorphism $\mathbb{I}_\mathbb{Q} \to G$. Deduce that $\mathbb{I}_\mathbb{Q}$ is isomorphic to the multiplicative group of positive rationals.

4.7.2. Let $\mathscr{I}$ be a fractional ideal in F.

(a) Show that $D_\mathscr{I} = \{\alpha \in \mathcal{O} : \alpha\mathscr{I} \subseteq \mathcal{O}\}$ is an ideal of $\mathcal{O}$.

(b) Let $F = \mathbb{Q}[\sqrt{-5}]$ and $\mathscr{I} = \mathbb{Z}[\sqrt{-5}] \cdot \frac{1+\sqrt{-5}}{3}$. Find $D_\mathscr{I}$.

4.7.3. Prove the statement about the norm in Prop. 4.7.6.

4.7.4. Let $\mathscr{I}$ be a fractional ideal in F. Prove that $\mathscr{I}^{-1} = \{\alpha \in F : \alpha\mathscr{I} \subseteq \mathcal{O}\}$.

4.7.5. Let $\mathscr{I}$ and $\mathscr{J}$ be two fractional ideals in F. Prove that $\mathscr{I} \subseteq \mathscr{J}$ if and only if $\mathscr{I}^{-1} \supseteq \mathscr{J}^{-1}$.

4.7.6. When do two triples (q, a, b) and (q', a', b') as in Cor. 4.7.5 give the same fractional ideal?

4.7.7. Show that $\mathcal{O}$ is a PID if and only if all fractional ideals are principal.

Orders

4.7.8. Let G be an abelian group and $M \subseteq G$ a multiplicatively closed subset. Show that $\langle M \rangle = \{gh^{-1} : g, h \in M\}$ is the smallest subgroup of G containing M. Let $M' \subseteq G'$ is another such pair, and $f : M \to M'$ a multiplication-preserving bijection. Show that f extends uniquely to a group isomorphism $\langle M \rangle \cong \langle M' \rangle$.

4.7.9. Let $\mathcal{O}_c \subseteq \mathcal{O}$ be the order of conductor c. Consider the sets

$$\mathbb{I}_c(c) = \{I_1 I_2^{-1} : I_1, I_2 \text{ ideals of } \mathcal{O}_c, \gcd(\mathrm{N}_c I_1, c) = \gcd(\mathrm{N}_c I_2, c) = 1\},$$
$$\mathbb{I}_1(c) = \{J_1 J_2^{-1} : J_1, J_2 \text{ ideals of } \mathcal{O}, \gcd(\mathrm{N} J_1, c) = \gcd(\mathrm{N} J_2, c) = 1\}.$$

(a) Show that

$$\mathbb{I}_c(c) = \{\tfrac{1}{e}I : e \in \mathbb{N}, I \text{ ideal of } \mathcal{O}_c \text{ with } \gcd(e, c) = \gcd(\mathrm{N}_c I) = 1\},$$

and similarly for $\mathbb{I}_1(c)$.

(b) Show that $\mathbb{I}_1(c)$ and $\mathbb{I}_c(c)$ are the smallest groups containing, respectively, $\mathbb{I}_1(c)^+$ and $\mathbb{I}_c(c)^+$ of Exer. 4.6.13.

(c) Show that the bijections between $\mathbb{I}_1(c)^+$ and $\mathbb{I}_c(c)^+$ of Exer. 4.6.13 extend to group isomorphisms $\mathbb{I}_1(c) \cong \mathbb{I}_c(c)$

It's often convenient to replace $\mathbb{I}_c(c)$ with the isomorphic group $\mathbb{I}_1(c)$. As c varies, the $\mathbb{I}_1(c)$ are all subgroups of $\mathbb{I}_F$.

4.8 Unique Factorization of Ideals

We are finally ready to prove unique factorization for ideals.

Theorem 1.5.5. *Let $F = \mathbb{Q}[\sqrt{D}]$ be a quadratic field and $\mathcal{O}$ its ring of integers. For any ideal I of $\mathcal{O}$, other than 0 and $\mathcal{O}$, there exist prime ideals $P_1, P_2, \ldots P_n$ of $\mathcal{O}$, not necessarily distinct, such that $I = P_1 P_2 \ldots P_n$. This factorization is unique up to permutation of factors.*

Proof. Existence: Let S be the set of all ideals $0 \subsetneq I \subsetneq \mathcal{O}$ without a prime factorization. Suppose that S were nonempty. By Prop. 4.4.3, $\mathcal{O}$ is a Noether ring, so Cor. 4.4.5 produces a largest element $L \in S$. The ideal L can't be prime, as each prime ideal is its own factorization. Then L is not maximal either, so L is strictly contained in a maximal ideal P by Cor. 4.4.6: $L \subsetneq P \subsetneq \mathcal{O}$.

Multiplication of fractional ideals preserves strict inclusion: if $J \subsetneq K$ but $IJ = IK$, cancelling I gives a contradiction $J = K$. Thus, multiplying $L \subsetneq P$ by P^{-1} gives $LP^{-1} \subsetneq \mathcal{O}$. Similarly, $P \subsetneq \mathcal{O}$ implies $\mathcal{O} \subsetneq P^{-1}$ (Exer. 4.7.5), which multiplied by L gives $L \subsetneq LP^{-1}$.

To summarize, $LP^{-1} \neq \mathcal{O}$ is an integral ideal strictly larger than L, and therefore not in S. Thus we have a prime factorization $LP^{-1} = P_1 \cdots P_k$. But then $L = P_1 \cdots P_k P$ is a prime factorization of L, contradicting $L \in S$. We see that S must be empty: each nontrivial ideal of $\mathcal{O}$ has a prime factorization.

Uniqueness: Let $I = P_1 \cdots P_r = Q_1 \cdots Q_s$. Then $P_1 \mid Q_1 \cdots Q_s$. By the ideal version of Euclid's Lemma (Exer. 2.5.1) we have that P_1 divides, say, Q_1 $(P_1 \supseteq Q_1)$. By Cor. 2.5.8 all nonzero prime ideals in $\mathcal{O}$ are maximal, so $P_1 = Q_1$. By Cor. 4.6.9, we can cancel $P_1 = Q_1$ to get $P_2 \cdots P_r = Q_2 \cdots Q_s$. Proceeding inductively, we match every P_i with some Q_j, so that one prime factorization of I is a permutation of the other.   ∎

Exercises

4.8.1. Let P be a nonzero prime ideal of $\mathcal{O}$, and $k \in \mathbb{N}$. By Exer. 2.5.9, $R = \mathcal{O}/P^k$ is a local ring. For $\alpha \in \mathcal{O}$ we write $\tilde{\alpha} = \alpha + P^k \in R$, and similarly for subsets of $\mathcal{O}$.

(a) Prove that $R \supseteq \tilde{P} \supseteq \tilde{P}^2 \supseteq \cdots \supseteq \tilde{P}^{k-1} \supseteq \tilde{P}^k = 0$ is the complete list of ideals of R. Show that they are all distinct.
(b) Let $0 \leq i < k$. Show that $\tilde{P}^i = R\tilde{\alpha}$ for any $\tilde{\alpha} \in \tilde{P}^i \setminus \tilde{P}^{i+1}$.
(c) Construct an isomorphism of additive groups $R/\tilde{P} \cong \tilde{P}^i/\tilde{P}^{i-1}$.
(d) Use Exer. 2.5.10 to prove that $R^\times \cong (1 + \tilde{P}) \times (\mathcal{O}/P)^\times$.

4.8.2. We generalize the Euler phi function to ideals $0 \subsetneq I \subsetneq \mathcal{O}$ by putting $\varphi(I) = |(\mathcal{O}/I)^\times|$. Prove the following:

(a) $\varphi(IJ) = \varphi(I)\varphi(J)$, for relatively prime ideals I and J of $\mathcal{O}$.
(b) $\varphi(P^k) = (NP)^{k-1}(NP - 1)$, for P a nonzero prime ideal of $\mathcal{O}$.

These two properties, along with Unique Factorization of Ideals, allow us to compute $\varphi(I)$ for any I.

4.9 Prime Ideals in $\mathcal{O}$

Before we start factoring ideals, we need an explicit description of prime ideals.

4.9.1 Lemma. *Let P be a nonzero prime ideal of $\mathcal{O}$. There exists a unique prime number $p \in \mathbb{N}$ such that $P \mid p$.*

Proof. Let $NP = pp' \ldots$ be the prime factorization in $\mathbb{N}$ of NP. The ideal P is prime, and divides $P\bar{P} = \mathcal{O} \cdot NP = (\mathcal{O}p)(\mathcal{O}p') \ldots$. By Euclid's Lemma for ideals, Exer. 2.5.1, we must have $P|p$, up to renaming. Then $NP \mid Np = p^2$, so $p \in \mathbb{N}$ must be the sole prime factor of NP. ∎

To find all nonzero prime ideals of $\mathcal{O}$, it's enough to factor the ideal $\mathcal{O}p$ for every prime number $p \in \mathbb{N}$. Let $\mathcal{O}p = P_1 P_2 \ldots$ be that factorization. Taking the norm, we get

$$p^2 = N(\mathcal{O}p) = (NP_1)(NP_2)\ldots,$$

which leaves only three possibilities.

4.9.2 Proposition-Definition. *Let $p \in \mathbb{N}$ be a prime. The prime factorization of $\mathcal{O}p$ has one of the following forms:*

(a) $\mathcal{O}p = P$ with $NP = p^2$; in other words, p remains prime when viewed as a principal ideal of $\mathcal{O}$. We say that p is **inert**.
(b) $\mathcal{O}p = P^2$ with $NP = p$. We say that p is **ramified**.
(c) $\mathcal{O}p = P\bar{P}$ with $NP = p$ and $P \neq \bar{P}$. We say that p is **split**.

The terms *inert, ramified* and *split* also apply to the prime ideal factors P (and $\bar{P}$) of $\mathcal{O}p$.

Proof. Let P be a prime ideal factor of $\mathcal{O}p$. If $NP = p^2$, we're in case (a). Otherwise, $\mathcal{O}p = \mathcal{O} \cdot NP = P\bar{P}$ must be the unique prime factorization of $\mathcal{O}p$. If $P = \bar{P}$, we're in case (b); otherwise, we're in case (c). ∎

4.9.3 Proposition. *Let $p \in \mathbb{N}$ be a prime, and let $\nu = 0, 1$ or 2 be the number of distinct solutions in $\mathbb{Z}/p\mathbb{Z}$ to the equation $x^2 - tx + n = 0$. Then the type of prime factorization of $\mathcal{O}p$ is determined by ν:*

(a) $\nu = 0$ if and only if p is inert.
(b) $\nu = 1$ if and only if p is ramified.
(c) $\nu = 2$ if and only if p is split.

Proof. Since a quadratic equation has at most two solutions in the field $\mathbb{Z}/p\mathbb{Z}$, it suffices to prove only the "if" direction in all three cases.

(a) Suppose that p is inert and that there exists an $r \in \mathbb{Z}$ such that $r^2 - tr + n \equiv 0 \pmod{p}$. Then $(r - \delta)(r - \bar{\delta}) = r^2 - tr + n \in \mathcal{O}p$. The ideal $\mathcal{O}p$ is prime, so it must contain one of the factors on the left, say $r - \delta$. But then $r/p - \delta/p \in \mathcal{O}$, which is a contradiction.

(b) and (c) Assume now that $\mathcal{O}p = P\bar{P}$, which describes both the ramified and the split case, according to whether $P = \bar{P}$ or not.

Let $P = d(\mathbb{Z}a + \mathbb{Z}(-b + \delta))$ be a standard form of P. Then $p = \mathrm{N}P = d^2 a$, so $d = 1, a = p$, and

$$P = \mathbb{Z}p + \mathbb{Z}(-b + \delta),$$
$$\bar{P} = \mathbb{Z}p + \mathbb{Z}(-b + \bar{\delta}) = \mathbb{Z}p + \mathbb{Z}(b - t + \delta).$$

As $\bar{\delta} - (t - b) = -(t - \bar{\delta}) + b = -\delta + b \in P$, we have

$$x^2 - tx + n = (x - \delta)(x - \bar{\delta}) \equiv (x - b)(x - (t - b)) \pmod{P}.$$

Since b and $t - b$ are both in $\mathbb{Z}$, modulo P they both end up in $\mathbb{Z}/p\mathbb{Z}$. We have that $\nu = 1$ if and only if $b \equiv t - b \pmod{p}$, and $\nu = 2$ otherwise.

The last congruence is also necessary and sufficient for P to be ramified. Indeed, the condition $P = \bar{P}$ is equivalent to the weaker one $P \subseteq \bar{P}$, or

$$(\mathbb{Z}p + \mathbb{Z}(-b + \delta)) \subseteq (\mathbb{Z}p + \mathbb{Z}(b - t + \delta)).$$

This, in turn, is equivalent to the existence of $k, l \in \mathbb{Z}$ such that $-b + \delta = kp + l(b - t + \delta)$. Comparing coefficients, we see this happens if and only if $l = 1, -b = kp + b - t$, or, equivalently, $b \equiv t - b \pmod{p}$.   ∎

4.9.4 Proposition. *The number of solutions to the equation $x^2 - tx + n = 0$ in $\mathbb{Z}/p\mathbb{Z}$ is $\nu = 1 + \left(\frac{D_F}{p}\right)$.*

For the definition of $\left(\frac{a}{2}\right)$, see Exer. 1.1.13.

Proof. Assume first that $p > 2$, so 2 is invertible in $\mathbb{Z}/p\mathbb{Z}$. This allows us to divide by 2 when completing the square, as in the standard proof of the quadratic formula:

$$0 = x^2 - tx + n = \left(x - \tfrac{t}{2}\right)^2 - \tfrac{t^2 - 4n}{4},$$

or, equivalently $(2x - t)^2 = D_F$. The number of solutions ν thus depends on whether D_F is a square modulo p, which is detected by the Legendre symbol of Def. 1.1.7:

$$\nu = \begin{cases} 0 \text{ if } D_F \text{ is a nonsquare in } \mathbb{Z}/p\mathbb{Z} \\ 1 \text{ if } D_F = 0 \pmod{p} \\ 2 \text{ if } D_F \text{ is a nonzero square in } \mathbb{Z}/p\mathbb{Z} \end{cases} = 1 + \left(\frac{D_F}{p}\right).$$

When $p = 2$, we can no longer complete the square. Instead, in Exer. 4.9.4 you will directly examine the four possible quadratic equations with coefficients in $\mathbb{Z}/2\mathbb{Z}$.   ∎

We summarize the previous results in the definitive statement on the factorization in $\mathcal{O}$ of integer primes.

4.9.5 Theorem. *Let* $p \in \mathbb{N}$ *be a prime.*

(a) *If* $\left(\frac{D_F}{p}\right) = -1$, $\mathcal{O}p$ *is a prime ideal in* $\mathcal{O}$: *p is inert.*

(b) *If* $\left(\frac{D_F}{p}\right) = 0$ *or* 1, *pick* $b \in \mathbb{Z}$ *so that* $b^2 - tb + n \equiv 0 \pmod{p}$. *Put* $P = \mathbb{Z}p + \mathbb{Z}(-b + \delta)$. *Then the following hold:*

- P *is a prime ideal in* $\mathcal{O}$.
- $\bar{P} = \mathbb{Z}p + \mathbb{Z}(-(t - b) + \delta)$.
- *We have the factorization into prime ideals* $\mathcal{O}p = P\bar{P}$.
- $\left(\frac{D_F}{p}\right) = 0$ *if and only if* $P = \bar{P}$, *or equivalently* $b \equiv t - b \pmod{p}$: *p is ramified.*
- $\left(\frac{D_F}{p}\right) = 1$ *if and only if* $P \neq \bar{P}$, *or equivalently* $b \not\equiv t - b \pmod{p}$: *p is split.*

Since $\left(\frac{D_F}{p}\right) = 0$ if and only if $p \mid D_F$, we have the following important corollary.

4.9.6 Corollary. *A prime* $p \in \mathbb{N}$ *is ramified if and only if* $p \mid D_F$. *In particular, only finitely many primes in* $\mathbb{N}$ *ramify.*

4.9.7 Example. Let $F = \mathbb{Q}[\sqrt{-14}]$, so $\mathcal{O} = \mathbb{Z}[\sqrt{-14}]$. The only ramified primes are 2 and 7, the prime divisors of $D_F = -56$. The element $\delta = \sqrt{-14}$ is a root of the equation $x^2 + 14 = 0$, which reduces to $x^2 = 0$ modulo both 2 and 7. We can then take $b = t - b = 0$ in Thm. 4.9.5 (b). We find that $\mathcal{O} \cdot 2 = P_2^2$, $\mathcal{O} \cdot 7 = P_7^2$ where

$$P_2 = \mathbb{Z} \cdot 2 + \mathbb{Z}\sqrt{-14} \text{ and } P_7 = \mathbb{Z} \cdot 7 + \mathbb{Z}\sqrt{-14}$$

are prime ideals of $\mathcal{O}$.

Let's consider the factorizations of 11 and 23. By Quadratic Reciprocity,

$$\left(\frac{-56}{11}\right) = \left(\frac{-1}{11}\right) = -1, \quad \left(\frac{-56}{23}\right) = \left(\frac{-14}{23}\right) = \left(\frac{9}{23}\right) = 1,$$

so that 11 is inert and 23 split. In fact $(\pm 3)^2 \equiv -14 \pmod{23}$, and the recipe in Thm. 4.9.5 (b) gives the prime factorization

$$\mathcal{O} \cdot 23 = (\mathbb{Z} \cdot 23 + \mathbb{Z}(-3 + \sqrt{-14}))(\mathbb{Z} \cdot 23 + \mathbb{Z}(3 + \sqrt{-14})). \qquad \square$$

4.9.8 Example. Let $F = \mathbb{Q}[\sqrt{-15}]$. We find that $D_F = -15$ and $\mathcal{O} = \mathbb{Z}[\delta]$, where $\delta = (1 + \sqrt{-15})/2$ satisfies $\delta^2 - \delta + 4 = 0$.

Let's factor $I = \mathbb{Z}\cdot 84 + \mathbb{Z}(49 + 7\delta)$ into prime ideals. We put I into standard form by factoring out $\gcd(84, 49, 7) = 7$:

$$(4.9.9) \qquad\qquad I = 7(\mathbb{Z}\cdot 12 + \mathbb{Z}(7 + \delta)).$$

In our general notation, $d = 7, a = 12$ and $b = -7$. Since $(-7)^2 - (-7) + 4 = 60 \equiv 0 \pmod{12}$, I is indeed an ideal. Its norm is $NI = d^2 a = 2^2 \cdot 3 \cdot 7^2$ by Exer. 4.6.3. We first figure out the factorizations of 2, 3 and 7 into prime ideals in $\mathcal{O}$:

2: Since $x^2 - x + 4 \equiv x^2 + x \equiv x(x + 1) \pmod 2$, the prime 2 splits: $\mathcal{O}\cdot 2 = P_2 \bar{P}_2 = (\mathbb{Z}\cdot 2 + \mathbb{Z}\delta)(\mathbb{Z}\cdot 2 + \mathbb{Z}(1 + \delta))$.
3: This is a divisor of D_F, so it must ramify. We find that $x^2 - x + 4 \equiv (x+1)^2$ $\pmod 3$, so $\mathcal{O}\cdot 3 = P_3^2$ with $P_3 = \mathbb{Z}\cdot 3 + \mathbb{Z}(1 + \delta)$.
7: As $\left(\frac{-15}{7}\right) = \left(\frac{6}{7}\right) = -1$, the principal ideal $\mathcal{O}\cdot 7$ is prime. Consequently, there is no ideal of norm 7.

We will factor I by finding sufficient powers of these prime ideals to account for all prime powers dividing NI in $\mathbb{N}$. The factor of 7 in (4.9.9) has norm $N(\mathcal{O}\cdot 7) = 7^2$, and so contributes all the powers of 7 in NI. Similarly, as $3 \mid NI$ and P_3 is the only ideal of norm 3, P_3 must divide I.

To account for the factor of 2^2 in NI, the ideal I must have two (possibly equal) prime ideal factors of norm 2. The product of the two can be $P_2^2, \bar{P}_2^2$, or $P_2\bar{P}_2 = \mathcal{O}\cdot 2$. We rule out the last possibility, since it's obvious from (4.9.9) that $2 \nmid I$. To decide whether $P_2^2 \mid I$ or $\bar{P}_2^2 \mid I$, we could use brute force: calculate $I(P_2^2)^{-1} = \frac{1}{4}I\bar{P}_2^2$ and $I(\bar{P}_2^2)^{-1} = \frac{1}{4}IP_2^2$, and see which is a subset of $\mathcal{O}$.

We can, however, avoid the tedious ideal multiplication altogether. Since $12 \in P_2$, we have that $P_2 \mid I$, i.e., $P_2 \supseteq I$, if and only if $7 + \delta \in P_2$, and similarly for $\bar{P}_2$. We calculate $7 + \delta = 3\cdot 2 + (1 + \delta) \in \bar{P}_2$, so that $\bar{P}_2 \mid I$. We finally get the prime factorization $I = 7\cdot P_3 \cdot \bar{P}_2^2$.   □

This example generalizes to a factorization algorithm for ideals with $d = 1$. In combination with Thm. 4.9.5, it allows us to explicitly factor any ideal.

4.9.10 Proposition. Let $I = \mathbb{Z}a + \mathbb{Z}(-b + \delta)$ be an ideal of $\mathcal{O}$. Let $a = p_1^{e_1}\cdots p_r^{e_r}$ be the prime factorization of a in $\mathbb{Z}$. The factorization of I into prime ideals is given by

$$I = \prod_{i=1}^{r}(\mathbb{Z}p_i + \mathbb{Z}(-b + \delta))^{e_i}.$$

Exercises

4.9.1. List, in standard form, all prime ideals in $\mathbb{Z}[\sqrt{-30}]$ of norm ≤ 50. Decide for each prime ideal whether it is inert, ramified, or split.

4.9.2. Factor the following ideals into primes in their respective rings of integers:

(a) $\mathbb{Z}[\sqrt{-17}] \cdot 30$
(b) $\mathbb{Z} \cdot 147 + \mathbb{Z}(42 + 7\sqrt{15})$
(c) (The principal ideal generated by) $5 + \sqrt{-5}$
(d) (The principal ideal generated by) $5 + 2\sqrt{85}$

4.9.3. Finish the computations of $I(P_2^2)^{-1}$ and $I(\bar{P}_2^{-2})^{-1}$ in Ex. 4.9.8.

4.9.4. We prove Prop. 4.9.4 when $p = 2$:

(a) Show that $D_F = t^2 - 4n \pmod 8$ depends only on $n \pmod 2$ and t $\pmod 4$. Write out the 2×4 table of the possible values of $D_F \pmod 8$.
(b) There are four different quadratic equations with coefficients in $\mathbb{Z}/2\mathbb{Z}$. Enter them in the first column of the table below, then fill in the four corresponding rows:

$x^2 - tx + n \equiv 0 \pmod 2$	ν	$D_F \pmod 8$	$\left(\frac{D_F}{2}\right)$

Here ν is the number of distinct solutions of the equation, and $\left(\frac{D_F}{2}\right)$ is defined in Exer. 1.1.13. Since the equations are determined by t, n $\pmod 2$, while $D_F \pmod 8$ depends on $t \pmod 4$, there may be several values in the third column corresponding to a single equation.

(c) Verify that $\nu = 1 + \left(\frac{D_F}{2}\right)$.

4.9.5.* Describe the prime factorization of an ideal I of $\mathcal{O}$ satisfying $\bar{I} = I$.

4.9.6. Let P be a ramified prime ideal of $\mathcal{O}$. Show that for all $\alpha \in \mathcal{O}$, $\bar{\alpha} \equiv \alpha$ $\pmod P$.

4.9.7. Let $c(\alpha) = \bar{\alpha}$ be the conjugation in F. By Prop. 4.1.3 e), $G = \{\mathrm{id}_F, c\}$ is the group of ring isomorphisms from F to F. Let $P \neq 0$ be a prime ideal of $\mathcal{O}$, and consider the chain of subgroups $G \supseteq D_P \supseteq E_P$ given by

$$D_P = \{\sigma \in G : \sigma(P) = P\}$$
$$E_P = \{\sigma \in G : \sigma\alpha \equiv \alpha \pmod P \text{ for all } \alpha \in F\}$$

(a) List all possible chains of subgroups $G \supseteq G_1 \supseteq G_2$ of length three.
(b) Show that the chain $G \supseteq D_P \supseteq E_P$ depends only on whether P is inert, ramified, or split. Do all the chains listed in (a) arise in this way?

4.9.8. Prove Prop. 4.9.10. Extend this to a recipe for factoring an arbitrary ideal of $\mathcal{O}$.

4.9.9. Let $r_1, \ldots, r_k, s_1, \ldots, s_l, i_1, \ldots, i_m \in \mathbb{N}$ be distinct positive primes. Prove that there exists a quadratic field in which all r_j are ramified, all s_j are split, and all i_j are inert.

4.9.10. By Exer. 2.2.16, every finite field of prime characteristic p has p^n elements. Galois theory shows that there is, up to isomorphism, only one such field for each n. We give an elementary proof for $n = 2$.

(a) Let K be a subfield of an arbitrary field L, with L of dimension 2 as a K-vector space (see Exer. 2.2.16). Show that any $\alpha \in L \setminus K$ is a root of some quadratic polynomial $x^2 + ax + b \in K[x]$.
(b) Construct a ring isomorphism $K[x]/\langle x^2 + ax + b \rangle \xrightarrow{\sim} L$. Deduce that $x^2 + ax + b$ is irreducible in $K[x]$.
(c) The derivation of the quadratic formula requires division by 2, which we can do in any field K of odd characteristic. In that case, show that $L \cong K[x]/\langle x^2 - m \rangle$ for some nonsquare $m \in K^\times$.
(d) Fix an odd prime $p \in \mathbb{N}$, and take $m, m' \in \mathbb{F}_p^\times \setminus \mathbb{F}_p^{\times 2}$. Construct a ring isomorphism $\mathbb{F}_p[x]/\langle x^2 - m \rangle \cong \mathbb{F}_p[x]/\langle x^2 - m' \rangle$. Conclude that any two fields with p^2 elements are isomorphic. We denote any one of them by $\mathbb{F}_{p^2}$.
(e) Construct $\mathbb{F}_4$ and show that it's unique.

4.9.11. Let $P \neq 0$ be a prime ideal of $\mathcal{O}$. Find the quotient $\mathcal{O}/P$ when P is inert, ramified, or split.

4.9.12. We will determine the ring structure of $\mathcal{O}/\mathcal{O}p$, for $p \in \mathbb{N}$ prime.

(a) Construct the appropriate ring isomorphism:
$$\mathcal{O}/\mathcal{O}p \cong \begin{cases} \mathbb{F}_{p^2}, & \text{for } p \text{ inert} \\ \mathbb{F}_p[x]/\langle x^2 \rangle, & \text{for } p \text{ ramified} \\ \mathbb{F}_p \times \mathbb{F}_p, & \text{for } p \text{ split} \end{cases}$$

(b) The unique ring homomorphism $\mathbb{F}_p = \mathbb{Z}/p\mathbb{Z} \to \mathcal{O}/\mathcal{O}p$ allows us to think of $\mathbb{F}_p$ as a subring of $\mathcal{O}/\mathcal{O}p$. In all three cases, describe the image of $\mathbb{F}_p$ and $(\mathcal{O}/\mathcal{O}p)^\times$ under the isomorphism of a).
(c) Show that the quotient $(\mathcal{O}/\mathcal{O}p)^\times/\mathbb{F}_p^\times$ is cyclic of order $p - \left(\frac{D_F}{p} \right)$. Exer. 2.3.11 is helpful.

Orders

4.9.13. Let $\mathcal{O}_c \subseteq \mathcal{O}$ be the order of conductor c. Show that the correspondence of Exer. 4.6.13 matches prime ideals of $\mathcal{O}_c$ in $\mathbb{I}_c(c)^+$ with prime ideals of $\mathcal{O}$ not dividing c.

Chapter 5
The Ideal Class Group and the Geometry of Numbers

5.1 The Ideal Class Group

It turns out that the group of fractional ideals $\mathbb{I}_F$ is not an interesting invariant of the quadratic field F: for different fields F, F', Exer. 5.1.7 shows that $\mathbb{I}_F \cong \mathbb{I}_{F'}$. To get an object which does reflect the arithmetic of F, we consider a quotient of $\mathbb{I}_F$.

5.1.1 Definition. *Let $\mathbb{P}_F$ be the subgroup of $\mathbb{I}_F$ consisting of all principal fractional ideals. The quotient*

$$\mathrm{Cl}(F) = \mathbb{I}_F / \mathbb{P}_F$$

is called the **ideal class group** *of F.*

The ideal class group is trivial precisely when all fractional ideals in F are principal, that is, when $\mathcal{O}$ is a principal ideal domain (PID). A nontrivial $\mathrm{Cl}(F)$ measures how far $\mathcal{O}$ is from being a PID. The second big theorem of this book states that it's never too far.

5.1.2 Theorem. *For any (quadratic) field F, $\mathrm{Cl}(F)$ is finite.*

We define the **class number** of F as $h(F) = |\mathrm{Cl}(F)|$. In Sec. 5.4 we'll get some hands-on experience with computing ideal class groups. We start, though, by thinking of $\mathrm{Cl}(F)$ as an abstract group quotient.

(a) Let $\mathscr{I}$ and $\mathscr{J}$ be fractional ideals. A typical element of $\mathrm{Cl}(F)$ is a coset $[\mathscr{I}] = \mathbb{P}_F \mathscr{I}$, which we call the **ideal class** of $\mathscr{I}$. By the definition of a quotient group, $[\mathscr{I}] = [\mathscr{J}]$ if and only if $\mathscr{I} = (\mathcal{O}\alpha)\mathscr{J} = \alpha \mathscr{J}$ for some $\alpha \in F \setminus 0$. Two ideals are in the same class precisely when they're proportional. In particular, the identity of $\mathrm{Cl}(F)$ is $[\mathcal{O}]$, the class of principal fractional ideals.

(b) For any fractional ideal $\mathscr{I}$ there exists an $k \in \mathbb{Z}$ such that $I = k\mathscr{I} \subseteq \mathcal{O}$. Then $[\mathscr{I}] = [k\mathscr{I}] = [I]$, so that every ideal class is represented by an

M. Trifković, *Algebraic Theory of Quadratic Numbers*, Universitext,
DOI 10.1007/978-1-4614-7717-4_5, © Springer Science+Business Media New York 2013

integral ideal. We can define ideal classes without leaving $\mathcal{O}$: $[I] = [J]$ for
ideals I, J of $\mathcal{O}$ if and only if there exist $\beta, \gamma \in \mathcal{O}$ such that $\gamma I = \beta J$.

(c) The inverse in $\mathrm{Cl}(F)$ is given by conjugation. The identity $I\bar{I} = \mathcal{O} \cdot \mathrm{N}I$
translates to $[I][\bar{I}] = [\mathcal{O}]$, or $[I]^{-1} = [\bar{I}]$.

(d) By Unique Factorization of Ideals, each element of $\mathrm{Cl}(F)$ is a product of
classes of prime ideals. In group-theoretic terms, those classes form a set
of generators for the group $\mathrm{Cl}(F)$. We obtain relations among them by
factoring principal ideals.

5.1.3 Example. We calculated in Exer. 4.9.2(c) that

$$\mathbb{Z}[\sqrt{-5}](5 + \sqrt{-5}) = P_2 \cdot P_3 \cdot P_5,$$

which gives us the relation $[P_2][P_3][P_5] = [\mathcal{O}]$ in $\mathrm{Cl}(\mathbb{Q}[\sqrt{-5}])$. The prime ideal
P_2 is not principal, by Exer. 2.3.6. The class $[P_2]$ has order 2, since $P_2^2 = \mathcal{O} \cdot 2$.
We will see later that $\mathrm{Cl}(\mathbb{Q}[\sqrt{-5}])$ is cyclic of order 2, generated by $[P_2]$. $\square$

We will deduce the finiteness of $\mathrm{Cl}(F)$ from a geometric property of ideals,
viewed as lattices in a plane.

5.1.4 Proposition. *Fix a positive constant $\mathfrak{M}_F$. Each of the following state-*
ments implies the next:

(a) Each fractional ideal $\mathscr{I}$ contains an $\alpha \neq 0$ for which $|\mathrm{N}\alpha| \leq \mathfrak{M}_F \cdot \mathrm{N}\mathscr{I}$.
(b) Each ideal class in $\mathrm{Cl}(F)$ contains an integral ideal I of norm $\mathrm{N}I \leq \mathfrak{M}_F$.
(c) The ideal class group of F is finite.

Proof. To show that (a) implies (b), we must find in every ideal class $[\mathscr{I}]$
an integral ideal of small norm. To do this, we look for $\alpha \in F \setminus 0$ such that
$I = \alpha \mathscr{I} \subseteq \mathcal{O}$ and $\mathrm{N}(\alpha \mathscr{I}) \leq \mathfrak{M}_F$, in other words

$$\alpha \in \mathscr{I}^{-1} \setminus 0 \text{ and } |\mathrm{N}\alpha| \leq \frac{\mathfrak{M}_F}{\mathrm{N}\mathscr{I}}.$$

Applying statement (a) to $\mathscr{I} = \mathscr{I}^{-1}$ produces such an α.

Now assume that (b) holds. To deduce (c), it suffices to show that there
are only finitely many nonzero ideals I of $\mathcal{O}$ with norm $\leq \mathfrak{M}_F$. For this, in
turn, it suffices to show that a fixed $B \in \mathbb{N}$ is the norm of only finitely many
ideals.

The standard form $I = d(\mathbb{Z}a + \mathbb{Z}(-b + \delta))$ is unique once we require
that $d > 0$ and $0 \leq b < a$. The equation $\mathrm{N}I = d^2 a = B$ has only finitely
many solutions $d, a \in \mathbb{N}$, and for each of them there are only finitely many b
satisfying $0 \leq b < a$ and $b^2 - tb + n \equiv 0 \pmod{a}$. ∎

Statement (a), and with it the finiteness of $\mathrm{Cl}(F)$, reduces to the purely
geometric task of finding a nonzero lattice point near the origin. A satisfying
sufficient condition for the existence of such a point is the subject of the next
section.

Exercises

5.1.1. Prove that $\mathbb{P}_F \cong F^\times/\mathcal{O}^\times$.

5.1.2. The order of a group element $g \in G$ is the smallest $n \in \mathbb{N}$ such that $g^n = 1$. In terms of the ideal I, when does the class $[I] \in \mathrm{Cl}(F)$ have order 2?

5.1.3. Determine the order of the class of each ideal below in the ideal class group of the corresponding field:

(a) $\mathbb{Z} \cdot 5 + \mathbb{Z}\sqrt{35}$
(b) $\mathbb{Z} \cdot 7 + \mathbb{Z}\sqrt{35}$
(c) $\mathbb{Z} \cdot 3 + \mathbb{Z}\frac{1+\sqrt{-23}}{2}$
(d) $\mathbb{Z} \cdot 13 + \mathbb{Z}(5 + i)$

5.1.4. Are the two ideals I and J from Exer. 5.1.3(a) and (b) in the same ideal class? If they are, find $\alpha \in \mathbb{Q}[\sqrt{35}]$ such that $J = \alpha I$.

5.1.5. We can define the ideal class group purely in terms of integral ideals:

(a) Let I and J be nonzero ideals of $\mathcal{O}$. Put $I \sim J$ if and only if there exist $\beta, \gamma \in \mathcal{O}$ for which $\gamma I = \beta J$. Show that $\sim$ is an equivalence relation on $\mathbb{I}_F^+$.
(b) Show that $[I][J] = [IJ]$ gives a well-defined operation on the quotient set $\mathrm{Cl}(F)_{\mathrm{int}} = \mathbb{I}_F^+/\sim$, which turns it into a group.
(c) Show that $\mathrm{Cl}(F)_{\mathrm{int}}$ coincides with $\mathrm{Cl}(F)$ of Def. 5.1.1.

5.1.6. Let F be a real quadratic field. We say that $\alpha \in F$ is **totally positive** if both $\alpha > 0$ and $\bar{\alpha} > 0$.

(a) Show that the set $F_{>0}$ of totally positive elements is a group under multiplication.
(b) Show that $N\alpha > 0$ if and only if α or $-\alpha$ is in $F_{>0}$.
(c) Let $\mathcal{O}_{>0}^\times = \mathcal{O}^\times \cap F_{>0}$ be the group of totally positive units. Show that

$$\mathcal{O}^\times/\mathcal{O}_{>0}^\times = \begin{cases} \mathbb{Z}/2\mathbb{Z} & \text{if } Nu = 1 \text{ for all } u \in \mathcal{O}^\times \\ \mathbb{Z}/2\mathbb{Z} \times \mathbb{Z}/2\mathbb{Z} & \text{if } Nu = -1 \text{ for some } u \in \mathcal{O}^\times. \end{cases}$$

(d) Find units $u, v \in \mathbb{Z}[\sqrt{2}]$ with $Nu = -1, Nv = 1$.
(e) Define the **narrow ideal class group** of F by $\mathrm{Cl}^+(F) = \mathbb{I}_F/\mathbb{P}_F^+$, where $\mathbb{P}_F^+$ is the group of principal fractional ideals with a totally positive generator. Construct a surjective group homomorphism $\mathrm{Cl}^+(F) \to \mathrm{Cl}(F)$, and determine its kernel.

5.1.7. By unique factorization, any nonzero ideal I of $\mathcal{O}$ can be written as a product ranging over all nonzero prime ideals:

$$I = \prod_P P^{e_P}.$$

Here $e_P \in \mathbb{Z}_{\geq 0}$, and $e_P = 0$ for all but finitely many P. For $I = \mathcal{O}$, we put $e_P = 0$ for all P. The set

$$\mathfrak{J} = \{(e_P) : e_P \in \mathbb{Z}, e_P = 0 \text{ for all but finitely many } P\}$$

is a group under componentwise addition.

(a) Show that the assignment $I \mapsto (e_P)$ extends to a group isomorphism $\mathbb{I}_F \cong \mathfrak{J}$.

(b) Conclude that for two fields F and F', $\mathbb{I}_F \cong \mathbb{I}_{F'}$.

5.1.8. Fix a bound $B \in \mathbb{R}_{>0}$. Prove, using unique factorization of ideals, that there are only finitely many ideals I of $\mathcal{O}$ with $NI \leq B$.

5.1.9. Let $\mathscr{I}$ be a fractional ideal in F, and $d \in \mathbb{N}$. Prove that there exists an ideal $I \subseteq \mathcal{O}$ in the class $[\mathscr{I}]$, for which $\gcd(NI, d) = 1$.

Orders

Let $\mathcal{O}_c \subseteq \mathcal{O}$ be the order of conductor c. In Exer. 4.7.9 we defined the group $\mathbb{I}_c(c)$ of fractional ideals of $\mathcal{O}_c$, and identified it with the subgroup $\mathbb{I}_1(c)$ of $\mathbb{I}_F$. To define the ideal class group of $\mathcal{O}_c$, we have two corresponding groups of principal fractional ideals:

$$\mathbb{P}_c(c) = \{\mathcal{O}_c \gamma : \gamma \in F_c^\times(c)\} \text{ and } \mathbb{P}_1(c) = \{\mathcal{O}\gamma : \gamma \in F_c^\times(c)\}.$$

Here we put

$$F_c^\times(c) = \{\alpha/\beta : \alpha, \beta \in \mathcal{O}_c, \gcd(N\alpha, c) = \gcd(N\beta, c) = 1\},$$
$$F_1^\times(c) = \{\alpha/\beta : \alpha, \beta \in \mathcal{O}, \gcd(N\alpha, c) = \gcd(N\beta, c) = 1\}.$$

Note that $\mathbb{P}_c(c)$ (resp. $\mathbb{P}_1(c)$) is the smallest subgroup of $\mathbb{I}_c(c)$ (resp. $\mathbb{I}_1(c)$) containing the principal ideals of $\mathcal{O}_c$ (resp. $\mathcal{O}$) with a generator that is in $\mathcal{O}_c$ *in both cases*. The isomorphism $\mathbb{I}_c(c) \cong \mathbb{I}_1(c)$ sending $\mathscr{I}$ to $\mathscr{I}\mathcal{O}$ identifies $\mathbb{P}_c(c)$ with $\mathbb{P}_1(c)$. The **ideal class group of the order** $\mathcal{O}_c$ is the quotient $\mathrm{Cl}(\mathcal{O}_c) = \mathbb{I}_c(c)/\mathbb{P}_c(c) \cong \mathbb{I}_1(c)/\mathbb{P}_1(c)$.

5.1.10. Show that $\mathbb{P}_1(c) \cdot \mathscr{I} \mapsto \mathbb{P}_F \cdot \mathscr{I}$ defines a surjective group homomorphism $\psi : \mathrm{Cl}(\mathcal{O}_c) \to \mathrm{Cl}(F)$ with kernel $(\mathbb{I}_1(c) \cap \mathbb{P}_F)/\mathbb{P}_1(c)$.

5.1.11. Construct group isomorphisms $\mathbb{I}_1(c) \cap \mathbb{P}_F \cong F_1^\times(c)/\mathcal{O}^\times$ and $\mathbb{P}_1(c) \cong F_c^\times(c)/(F_c^\times(c) \cap \mathcal{O}^\times) \cong F_c^\times(c)\mathcal{O}^\times/\mathcal{O}^\times \cong F_c^\times(c)/\mathcal{O}_c^\times$.

5.1.12. Let $\psi : \mathcal{O} \to \mathcal{O}/c\mathcal{O}$ be the natural ring homomorphism. Reducing the inclusion $\mathbb{Z} \subset \mathcal{O}$ modulo c allows to view $\mathbb{Z}/c\mathbb{Z}$ as a subring of $\mathcal{O}/c\mathcal{O}$.

(a) Show that $\mathcal{O}_c = \psi^{-1}(\mathbb{Z}/c\mathbb{Z})$.

(b) Extend ψ to a multiplicative homomorphism $\psi^\times : F_1^\times(c) \to (\mathcal{O}/c\mathcal{O})^\times$. Show that $\psi^\times$ is onto, and that $F_c^\times(c) = (\psi^\times)^{-1}((\mathbb{Z}/c\mathbb{Z})^\times)$.

(c) Conclude that the composed homomorphism $\varphi : F_1^\times(c) \to (\mathcal{O}/c\mathcal{O})^\times \to$ $(\mathcal{O}/c\mathcal{O})^\times/(\mathbb{Z}/c\mathbb{Z})^\times$ induces an isomorphism $F_1^\times(c)/F_c^\times(c) \cong (\mathcal{O}/c\mathcal{O})^\times/$ $(\mathbb{Z}/c\mathbb{Z})^\times$.

5.1.13. Using the preceding exercises and the Second and Third Isomorphism Theorems for abelian groups, verify the following isomorphisms:

$$\ker\psi = (\mathbb{I}_1(c) \cap \mathbb{P}_F)/\mathbb{P}_1(c) \cong (F_1^\times(c)/\mathcal{O}^\times)/(F_c^\times(c)\mathcal{O}^\times/\mathcal{O}^\times)$$
$$\cong F_1^\times(c)/(F_c^\times(c)\mathcal{O}^\times) \cong [F_1^\times(c)/F_c^\times(c)]/[F_c^\times(c)\mathcal{O}^\times/F_c^\times(c)]$$
$$\cong [(\mathcal{O}/c\mathcal{O})^\times/(\mathbb{Z}/c\mathbb{Z})^\times]/\varphi(\mathcal{O}^\times).$$

The last group is finite and easily computable, and measures how much bigger $\mathrm{Cl}(\mathcal{O}_c)$ is than $\mathrm{Cl}(F)$.

5.1.14. Calculate the group structure of $[(\mathcal{O}/c\mathcal{O})^\times/(\mathbb{Z}/c\mathbb{Z})^\times]/\varphi(\mathcal{O}^\times)$ for orders of discriminant $D = c^2 D_F = \{-100, -180, 180\}$.

5.2 Minkowski's Theorem

The geometric fact alluded to at the end of the previous section is Minkowski's theorem: a planar region that is sufficiently big, symmetric, and regular contains a nonzero lattice point.

5.2.1 Definition. *We say that a set $S \subseteq \mathbb{R}^2$ is* **nice** *if it satisfies all of the following:*

(a) S is centrally symmetric around 0: if $x \in S$, then $-x \in S$.
(b) S is convex: the line segment joining any two points in S is entirely inside S.
(c) S is measurable: we can make sense of the integral $A(S) = \iint\limits_S \mathrm{d}x\mathrm{d}y$
  defining the area of S.

We won't worry about the technical condition (c), as it will be obviously satisfied by all the S we encounter. The pictures in Fig. 5.1(a) and (b) are two examples of nice regions. We denote by $A(\Lambda)$ the area of the fundamental parallelogram of a lattice Λ.

5.2.2 Theorem (Minkowski). *Let $\Lambda, S \subseteq \mathbb{R}^2$ be a lattice and a nice region, respectively. If $A(S) > 4A(\Lambda)$, then there exists a nonzero point in $S \cap \Lambda$. If S is closed, the weaker condition $A(S) \geq 4A(\Lambda)$ suffices.*

The region in Fig. 5.1(a) satisfies all conditions of the theorem and indeed contains a nonzero lattice point. The remaining regions contain no lattice points except 0, even though all have area $\geq 4A(\Lambda)$. Each fails one of the requirements of the theorem:

(b) The region is not closed because it doesn't contain its boundary (the dashed line). Even though it is nice and its area is precisely $4A(\Lambda)$,

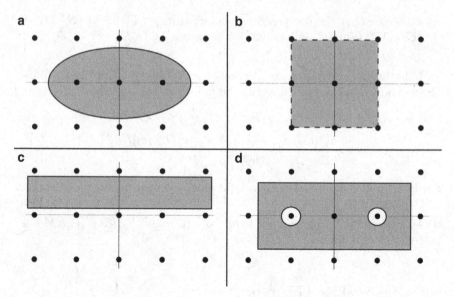

Fig. 5.1 Examples of regions in $\mathbb{R}^2$

it contains no lattice points except 0. For nonclosed regions, the strict
inequality in Thm. 5.2.2 is necessary.

(c) The region is convex, but not centrally symmetric.
(d) The region is centrally symmetric but not convex, allowing us to avoid
lattice points.

The area condition of Minkowski's theorem is sufficient, but not necessary.
Any $x \in \Lambda$ can be joined to $-x$ by a nice rectangle of arbitrarily small area.

Proof of Minkowski's Thm. 5.2.2. Before giving the formal argument, let's
look at Figs. 5.2 and 5.3 to see what's going on. The elements of Λ are drawn
as dots, the bold ones belonging to the scaled-up lattice 2Λ with fundamental
parallelogram Π.

We have marked by A through F the six translates of Π that intersect the
nice region S. Imagine those parallelograms as tiles made of transparent glass
on which S is painted in solid color. Stack the tiles on top of one another,
as in Fig. 5.3, and shine a light vertically down on the stack. The solid
regions that make up S will cast a shadow on the bottom parallelogram,
which has area $A(\Pi) = 4A(\Lambda)$. The total area of the painted regions adds
up to $A(S)$. By assumption, this is greater than $4A(\Lambda)$, so some two tiles
must have overlapping shadows. Pick a point z in that overlap, and take two
points $x, y \in S$ above z lying on different tiles. Having the same shadow
means that x and y have the same relative position within their respective
tiles (see Fig. 5.2). We see from that figure, and will prove shortly, that the
point γ halfway between x and $-y$ is in both $\Lambda \setminus 0$ and S, as desired.

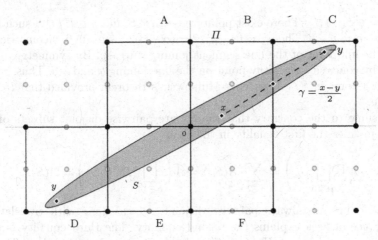

Fig. 5.2 A nice region S covered by parallelograms.

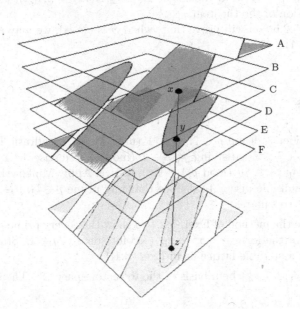

Fig. 5.3 The stack of tiles.

Formally, think of the bottom parallelogram in Fig. 5.3 as the fundamental parallelogram Π of 2Λ. For $\alpha \in 2\Lambda$, denote by $\Pi_\alpha = \{p + \alpha : p \in \Pi\}$ the translate of Π by α. The set $S_\alpha = (\Pi_\alpha \cap S) - \alpha \subseteq \Pi$ is the rigorous incarnation of the shadow cast by $\Pi_\alpha \cap S$ on the bottom parallelogram in Fig. 5.3. Grant for now the following claim:

(5.2.3)   There exist distinct $\alpha, \beta \in 2\Lambda$ for which $S_\alpha \cap S_\beta \neq \emptyset$.

Take $z \in S_\alpha \cap S_\beta$. There exist points $x \in \Pi_\alpha \cap S$ and $y \in \Pi_\beta \cap S$ such that $z = x - \alpha = y - \beta$. Then $\gamma = (x - y)/2 = (\alpha - \beta)/2$ is in $\Lambda \setminus 0$. Geometrically, γ is the midpoint of the line segment joining x to $-y$. By symmetry, $-y$ is in S; by concavity, so is any point on the line joining x and $-y$. Thus, γ is a nonzero point in $S \cap \Lambda$, proving Minkowski's theorem provided that (5.2.3) is true.

Assume to the contrary that the S_α are pairwise disjoint subsets of Π. That justifies the first equality in the chain

$$A \left(\bigsqcup_{\alpha \in 2\Lambda} S_\alpha \right) = \sum_{\alpha \in 2\Lambda} A(S_\alpha) = A \left(\bigsqcup_{\alpha \in 2\Lambda} (\Pi_\alpha \cap S) \right) = A(S).$$

The $\Pi_\alpha \cap S$ are always pairwise disjoint, which along with translation-invariance of area explains the second equality. The third equality is true since the translates of Π cover the whole plane, so in particular all of S. From $\bigsqcup_{\alpha \in 2\Lambda} S_\alpha \subseteq \Pi$ we deduce that $A(S) \leq A(\Pi) = 4A(\Lambda)$, contradicting the assumption of the theorem.

In Exer. 5.2.5 you will prove that, when S is closed, we may assume that $A(S) \geq 4A(\Lambda)$.   ∎

Exercises

5.2.1.* Consider a prime $p \in \mathbb{N}$, $p \equiv 1$ (mod 4). By Quadratic Reciprocity, we can find a $j \in \mathbb{Z}$ satisfying $j^2 \equiv -1$ (mod p). Exercise 3.1.1 shows that $\Lambda = \{ [\begin{smallmatrix} x \\ y \end{smallmatrix}] \in \Lambda_0 : x \equiv jy \pmod{p} \} \subset \mathbb{R}^2$ a lattice. Apply Minkowski's theorem to Λ and a suitable disc S to deduce that any prime $p \in \mathbb{N}$, $p \equiv 1$ (mod 4), is a sum of two squares.

5.2.2. Mimic the method of Exer. 5.2.1 to show that every prime $p \in \mathbb{N}, p \equiv 1$ (mod 3), is of the form $p = x^2 + xy + y^2$ for some $x, y \in \mathbb{Z}$. Start with the definition of a suitable lattice as in Exer. 3.1.1.

5.2.3. Let $\{v_1, \ldots, v_n\}$ be a basis of the $\mathbb{R}$-vector space $\mathbb{R}^n$. The subsets

$$\Lambda = \mathbb{Z}v_1 + \cdots + \mathbb{Z}v_n$$
$$\Pi = \{t_1 v_1 + \cdots + t_n v_n : t_i \in [0, 1) \text{ for all } 1 \leq i \leq n\}$$

are, respectively, the **rank** n **lattice** in $\mathbb{R}^n$ with basis $\{v_1, \ldots, v_n\}$, and the corresponding **fundamental parallelotope**. Prove the n-dimensional generalization of Minkowski's theorem:

Let $S \subset \mathbb{R}^n$ be a convex, measurable region, centrally symmetric about the origin. If $V(S) \geq 2^n V(\Pi)$, then S contains a nonzero point of Λ.

Here $V(T) = \int_T dx_1 \cdots dx_n$ is the standard n-dimensional volume of a measurable region T. In particular, $V(\Pi)$ is the absolute value of the determinant of the matrix with columns $v_1, \ldots, v_n$.

5.2.4. Here we prove a theorem of Legendre: each prime in $\mathbb{N}$ is a sum of four squares (some of which may be 0):

(a) Consider the subsets of $\mathbb{Z}/p\mathbb{Z}$ given by

$$K = \{k^2 : k \in \mathbb{Z}/p\mathbb{Z}\}, \quad L = \{-l^2 - 1 : l \in \mathbb{Z}/p\mathbb{Z}\}.$$

Count the number of elements in K and L to conclude that $p \mid k^2 + l^2 - 1$ for some $k, l \in \mathbb{Z}$.

(b) For the k and l you just found, put

$$\Lambda = \{(a, b, c, d) : \left[\begin{smallmatrix} c \\ d \end{smallmatrix}\right] \equiv \left[\begin{smallmatrix} k & l \\ -l & k \end{smallmatrix}\right] \left[\begin{smallmatrix} a \\ b \end{smallmatrix}\right] \pmod{p}\} \subset \mathbb{R}^4.$$

Prove that Λ is a rank 4 lattice by finding a basis. Show that its fundamental parallelotope has volume p^2.

(c) Find the smallest four-dimensional ball which must contain a point in $\Lambda \setminus 0$ by the generalized Minkowski's theorem of Exer. 5.2.3. Prove that $a^2 + b^2 + c^2 + d^2 \equiv 0 \pmod{p}$ for any $(a, b, c, d) \in \Lambda$. Deduce Legendre's theorem. (Note: The volume of a four-dimensional ball of radius r is $\pi^2 r^4 / 2$.)

5.2.5. Here we prove the last sentence in Minkowski's theorem: if the region S is closed, then there exists a point in $S \cap (\Lambda \setminus 0)$ even under the weaker condition $A(S) \geq 4A(\Lambda)$. We will show that the claim (5.2.3) still holds, so that the proof of Thm. 5.2.2 can proceed unchanged. To argue by contradiction, we assume that $A(S) = 4A(\Lambda)$ and that the S_α are pairwise disjoint.

(a) Prove that the S_α are closed subsets of Π, and that their union is not connected. Conclude that there exists a $w \in \Pi \setminus \sqcup_{\alpha \in 2\Lambda} S_\alpha$.

(b) Let $\overline{S_\alpha}$ be the closure of S_α in $\mathbb{R}^2$. It consists of S_α and the adjacent portion of the boundary of $\overline{\Pi}$. Show that the distance function $d_w : \cup \overline{S_\alpha} \to \mathbb{R}, d_w(x) = |x - w|$ has a nonzero minimum value r.

(c) Consider the disk of small enough radius around w to derive a contradiction with our assumption $A(S) = A(\Pi)$.

5.3 Application to Ideals

Fix a quadratic field F. In Prop. 5.1.4 we showed how to deduce the finiteness of the ideal class group $\mathrm{Cl}(F)$ from the statement

(5.3.1)   There exists a constant $\mathfrak{M}_F$ for which the following holds: each fractional ideal $\mathscr{I}$ contains an $\alpha \neq 0$ for which $|N\alpha| \leq \mathfrak{M}_F \cdot N\mathscr{I}$.

The value of $\mathfrak{M}_F$ depends only on the field F, not on the fractional ideal. To prove the finiteness of $\mathrm{Cl}(F)$ we need to find $\mathfrak{M}_F$ big enough for (5.3.1)

to hold. To efficiently compute ideal class groups, we'll use the bound in Prop. 5.1.4(b), for which we want $\mathfrak{M}_F$ to be as small as possible. We will determine such $\mathfrak{M}_F$ separately for real and imaginary quadratic fields.

Start with F an imaginary quadratic field. By Prop. 4.4.2, we may think of any fractional ideal $\mathscr{I}$ in F as a lattice in the complex plane.

5.3.2 Proposition. *Let $\mathscr{I}$ be a fractional ideal in an imaginary quadratic field F. The fundamental parallelogram of $\mathscr{I}$ has area $A(\mathscr{I}) = (\sqrt{|D_F|}/2) \cdot \mathrm{N}\mathscr{I}$.*

Proof. Write $\mathscr{I} = q(\mathbb{Z}a + \mathbb{Z}(-b + \delta))$ as in Cor. 4.7.5. Its fundamental parallelogram with respect to the basis $\{qa, q(-b + \delta)\}$ has base $|qa|$ and height $|q \operatorname{Im} \delta|$, hence $A(\mathscr{I}) = |q^2 a| \, |\operatorname{Im} \delta| = \mathrm{N}\mathscr{I}\sqrt{|D_F|}/2$. ∎

To apply Minkowski's theorem to the lattice $\mathscr{I}$ and a closed disc S centered at the origin, we need the disc to satisfy $A(S) \geq 4A(\mathscr{I}) = 2\sqrt{|D_F|} \cdot \mathrm{N}\mathscr{I}$. The smallest such disc is given by

$$S = \left\{ z : |z| \leq \sqrt{(2/\pi)\sqrt{|D_F|} \cdot \mathrm{N}\mathscr{I}} \right\}.$$

Minkowski's theorem produces a nonzero point $\alpha \in S \cap \mathscr{I}$, for which

$$(5.3.3) \qquad \mathrm{N}\alpha = \alpha\bar{\alpha} = |\alpha|^2 \leq \frac{2}{\pi}\sqrt{|D_F|} \cdot \mathrm{N}\mathscr{I}.$$

This is just (5.3.1) with $\mathfrak{M}_F = 2\sqrt{|D_F|}/\pi$. As in Prop. 5.1.4, we deduce that $\mathrm{Cl}(F)$ is finite when F is imaginary quadratic.

By contrast, the ring of integers $\mathcal{O}$ of a real quadratic field F is a dense subset of the real line. To apply Minkowski's theorem to $\mathcal{O}$ and its ideals, we need a way of viewing them as lattices in a plane.

When thinking of $\mathbb{Q}[\sqrt{319}]$ as a subset of $\mathbb{R}$, we're following the usual calculus convention by putting $\sqrt{319} = 17.8605\ldots$. Algebraically speaking, this is misleading: the only property of $\sqrt{319}$ we've ever used is that it's a solution to $x^2 - 319 = 0$. We lose information unless both solutions to that equation get equal billing. This suggests identifying $\sqrt{319}$ with the vector $(-17.8605\ldots, 17.8605\ldots) \in \mathbb{R}^2$, which extends to an arbitrary $\alpha \in \mathbb{Q}[\sqrt{319}]$ by

$$\rho(\alpha) = (\bar{\alpha}, \alpha)$$

This formula defines an injection $\rho : F \hookrightarrow \mathbb{R}^2$ for any real quadratic field F (see Fig. 5.4). We know that $\mathcal{O} = \mathbb{Z} + \mathbb{Z}\frac{D_F + \sqrt{D_F}}{2}$. Its embedding $\rho(\mathcal{O})$ is a lattice in $\mathbb{R}^2$ because it is spanned by the linearly independent vectors (Fig. 5.4)

$$\rho(1) = (1,1) \text{ and } \rho\left(\frac{D_F + \sqrt{D_F}}{2}\right) = \left(\frac{D_F - \sqrt{D_F}}{2}, \frac{D_F + \sqrt{D_F}}{2}\right).$$

Similarly, $\rho(\mathscr{I})$ is a lattice for any fractional ideal $\mathscr{I}$. It satisfies the following analog of Prop. 5.3.2.

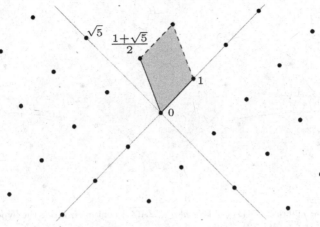

Fig. 5.4 The lattice $\rho\left(\mathbb{Z}\left[\frac{1+\sqrt{5}}{2}\right]\right)$ with a fundamental domain shaded in. The natural axes here are the rational axis $y = x$, and the irrational axis $y = -x$

5.3.4 Proposition. *Let $\mathscr{I}$ be a fractional ideal in a real quadratic field F. The fundamental parallelogram of $\rho(\mathscr{I})$ has area $A(\mathscr{I}) = \sqrt{D_F} \cdot N\mathscr{I}$.*

When F is imaginary, the distance-to-origin-squared function $\mathbb{C} \to \mathbb{R}, z \mapsto |z|^2$, restricts to the norm on F. In (5.3.3), this provided the crucial link between the geometry of the lattice $\mathscr{I}$ and the arithmetic of the field F. An analogous function $\mathbb{R}^2 \to \mathbb{R}$ in the real case is $(x, y) \mapsto xy$, which sends $\rho(\alpha) = (\bar{\alpha}, \alpha)$ to $\bar{\alpha}\alpha = N\alpha$.

Let $\mathscr{I}$ be a fractional ideal in a real quadratic field F. Consider the subsets of $\mathbb{R}^2$ depicted in Fig. 5.5 and defined by

$$H = \{(x, y) : |xy| \le A(\mathscr{I})/2\}$$
$$S = \{(x, y) : |x \pm y| \le \sqrt{2A(\mathscr{I})}\}.$$

It is natural to consider H, since we want to bound the norm of $\alpha \in \mathscr{I}$ by a multiple of $A(\mathscr{I})$. Unfortunately, Minkowski's theorem doesn't apply to the non-convex set H. Instead, we work with S, the biggest square contained in H. The bounds in the definition of H and S were chosen so that $A(S) = 4A(\mathscr{I})$. Applying Minkowski's theorem to S and the lattice $\rho(\mathscr{I})$ produces an $\alpha \in \mathscr{I} \setminus 0$ for which $\rho(\alpha) = (\bar{\alpha}, \alpha) \in S \subset H$. By definition of H,

$$|N\alpha| = |\alpha\bar{\alpha}| \le A(\mathscr{I})/2 = (\sqrt{D_F}/2) \cdot N\mathscr{I}.$$

This is the statement of (5.3.1) with $\mathfrak{M}_F = \sqrt{D_F}/2$. Proposition 5.1.4 now shows that $\text{Cl}(F)$ is finite for a real quadratic field F.

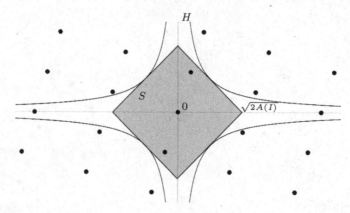

Fig. 5.5 The square $S = \{(x,y) : |x \pm y| \le \sqrt{2A(I)}\}$ contained inside the hyperbolic star $H = \{(x,y) : |xy| \le A(I)/2\}$.

Exercises

5.3.1. For the given ideal I in an imaginary quadratic field, list all $\alpha \in I$ satisfying the bound of (5.3.3):

(a) $\mathbb{Z} \cdot 14 + \mathbb{Z}(3 + \sqrt{-19})$
(b) $\mathbb{Z} \cdot 5 + \mathbb{Z}(2 - \sqrt{-26})$
(c) $\mathbb{Z} \cdot 47 + \mathbb{Z}\frac{47+\sqrt{-47}}{2}$

5.3.2. Let $\mathscr{I}$ be a fractional ideal in a real quadratic field F. Show that $\rho(\mathscr{I}) \subset \mathbb{R}^2$ is a lattice and that $A(\mathscr{I}) = \sqrt{D_F} \cdot \mathrm{N}\mathscr{I}$.

5.3.3. The complex conjugation restricted to any imaginary quadratic field gives the field conjugation. There is no such universal recipe for conjugating in a real quadratic F, which is why we need to remember both conjugates in defining the embedding $\rho : F \hookrightarrow \mathbb{R}^2$. To appreciate the difference, consider the decimal expansions

$$\alpha = 2.5 + i \cdot 3.122498\ldots \text{ and } \beta = -16.734993\ldots$$

Given that α and β are both quadratic numbers, try finding $\bar{\alpha}$ and $\bar{\beta}$.

5.3.4. We can sometimes use the embedding of a real quadratic field into $\mathbb{R}^2$ to prove the division algorithm, by analogy with Prop. 1.2.5.

(a) Prove that $\nu(\alpha) = |\mathrm{N}\alpha|$ defines a Euclid size on $\mathbb{Z}[\frac{1+\sqrt{13}}{2}]$. Do this by drawing an analog of Fig. 1.2 for the fundamental parallelogram of $\rho(\mathbb{Z}[\frac{1+\sqrt{13}}{2}]) \subset \mathbb{R}^2$. Caution: If $\nu(\beta - \alpha) < 1$, then $\rho(\beta)$ no longer has to lie inside the circle of radius 1 around $\rho(\alpha)$.
(b) Draw an analogous picture for $\mathbb{Z}[\sqrt{14}]$ to show that $|\mathrm{N}\alpha|$ is not a Euclid size on $\mathbb{Z}[\sqrt{14}]$. Nevertheless, $\mathbb{Z}[\sqrt{14}]$ is a Euclid domain for a different size function, as was proved only in 2004 by Harper [6].

5.3.5. Minkowski's theorem constructs a non-trivial unit in the ring of integers of a real quadratic field F. For that, we can think of a real quadratic number $\alpha = a + b\sqrt{D}$ as a real number by interpreting the symbol $\sqrt{D}$ as the positive square root. The absolute value of that real number is denoted $|\alpha|$.

(a) Fix $\alpha \in \mathcal{O} \setminus 0$. Find positive real constants A, B for which Minkowski's theorem applied to the lattice $\rho(\mathcal{O}) \subset \mathbb{R}^2$ and the rectangle

$$S = \{(x, y) : |x| < A, |y| < B\} \subseteq \mathbb{R}^2$$

produces a nonzero $\beta \in \mathcal{O}$ with $|N\beta| \leq \sqrt{D_F}$ and $|\beta| < |\alpha|$.

(b) Iterate the procedure of part (a) to get a sequence $\alpha_1, \alpha_2, \cdots \in \mathcal{O}$ with $|\alpha_1| > |\alpha_2| > \ldots$ and $N(\mathcal{O}\alpha_i) = |N\alpha_i| \leq \sqrt{D_F}$. By the proof of Prop. 5.1.4(c), there exist only finitely many possibilities for $\mathcal{O}\alpha_i$. Deduce that $\mathcal{O}$ has a unit other than ± 1.

In Ch. 6 we will explicitly compute the units in a real quadratic field.

5.4 Some Ideal Class Group Computations

We extract from the preceding section, and Prop. 5.1.4(b), the statement that will allow us to explicitly list elements in $\mathrm{Cl}(F)$.

5.4.1 Proposition (Minkowski's Bound). *Each ideal class in* $\mathrm{Cl}(F)$ *contains an ideal of norm at most* $\mathfrak{M}_F$, *with*

$$\mathfrak{M}_F = \sqrt{|D_F|} \cdot \begin{cases} \frac{2}{\pi} & \text{for } F \text{ imaginary} \\ \frac{1}{2} & \text{for } F \text{ real}. \end{cases}$$

To get a sense of how to use this proposition to compute ideal class groups, we begin with a few simple examples.

5.4.2 Example. Let $F = \mathbb{Q}[i]$, $\mathcal{O} = \mathbb{Z}[i]$. Minkowski's bound allows us to represent any ideal class by an ideal I satisfying

$$NI \leq \mathfrak{M}_F = \frac{2}{\pi}\sqrt{|-4|} \approx 1.27.$$

Since $NI \in \mathbb{Z}$, we must have $NI = 1$, so $I = \mathcal{O}$. Thus, each class contains $\mathcal{O}$, $\mathrm{Cl}(F) = \{[\mathcal{O}]\}$, and $\mathbb{Z}[i]$ is a PID. We have known this for a long time, of course; the point is that we can now prove it without resorting to the division algorithm in $\mathbb{Z}[i]$. □

5.4.3 Example. Let $F = \mathbb{Q}[\sqrt{-19}]$, so $\mathcal{O} = \mathbb{Z}[(1 + \sqrt{-19})/2]$. We will show that $\mathcal{O}$ is a PID, despite not being a Euclid domain by Exer. 2.3.9.

Minkowski's bound tells us that each ideal class has a representative I with

$$NI \leq \mathfrak{M}_F = \frac{2}{\pi}\sqrt{19} \approx 2.7.$$

If $NI = 1$, then $I = \mathcal{O}$, and $[I]$ is the identity in $\mathrm{Cl}(F)$. If $NI = 2$, then I is prime factor of 2. This is impossible: 2 is inert in $\mathbb{Q}[\sqrt{-19}]$, and the only prime ideal dividing it, namely $\mathcal{O} \cdot 2$ itself, has norm 4. We conclude that $h(F) = 1$. $\qquad\qquad\square$

5.4.4 Example. Let $F = \mathbb{Q}[\sqrt{-5}]$, so $\mathcal{O} = \mathbb{Z}[\sqrt{-5}]$ and $D_F = -20$. Let $P_2 = \mathbb{Z} \cdot 2 + \mathbb{Z}(1 + \sqrt{-5})$ be the unique ideal of norm 2. We saw in Ex. 5.1.3 that $\mathrm{Cl}(F)$ has at least two elements, $[\mathcal{O}]$ and $[P_2]$. There aren't any more: Minkowski's bound shows that each nonidentity ideal class contains an ideal I with

$$2 \leq NI \leq \frac{2}{\pi}\sqrt{20} \approx 2.8,$$

whence $NI = 2$ and $I = P_2$. Thus $\mathrm{Cl}(F)$ is cyclic of order 2, generated by $[P_2]$. $\qquad\qquad\square$

5.4.5 Example. Let $F = \mathbb{Q}[\sqrt{15}]$, so $\mathcal{O} = \mathbb{Z}[\sqrt{15}]$ and $D_F = 60$. By Minkowski's bound, each ideal class bar $[\mathcal{O}]$ contains an ideal I, whose norm is bounded by

$$2 \leq NI \leq \frac{1}{2}\sqrt{60} \approx 3.9,$$

that is, $NI = 2$ or 3. There exists a unique ideal of either norm, since 2 and 3 divide D_F and are therefore ramified:

$$(5.4.6) \qquad \begin{aligned} \mathcal{O} \cdot 2 &= P_2^2 \quad \text{where} \quad P_2 = \mathbb{Z} \cdot 2 + \mathbb{Z}(1 + \sqrt{15}) \\ \mathcal{O} \cdot 2 &= P_3^3 \quad \text{where} \quad P_3 = \mathbb{Z} \cdot 3 + \mathbb{Z}\sqrt{15}. \end{aligned}$$

Thus, $\mathrm{Cl}(F)$ has at most three elements, $[\mathcal{O}], [P_2]$, and $[P_3]$, with the understanding that some of the classes on the list may coincide. Moreover, (5.4.6) shows that $[P_2]$ is either trivial or has order 2, and ditto for $[P_3]$. To decide which, we use the following straightforward lemma.

5.4.7 Lemma. *Let F be a quadratic field with ring of integers $\mathcal{O}$. Let P be an ideal of $\mathcal{O}$ with prime norm. Then P is principal if and only if there exists an $\alpha \in \mathcal{O}$ with $N\alpha = \pm NP$.*

To apply the lemma to P_2 and P_3, we look for solutions in $\mathbb{Z}$ to $N(x + y\sqrt{15}) = x^2 - 15y^2 = \pm 2$ or ± 3. Reducing these equations modulo 5, we get $x^2 = \pm 2 \pmod 5$, which has no solutions, as $\left(\frac{\pm 2}{5}\right) = -1$. We conclude that $[P_2] \neq [\mathcal{O}] \neq [P_3]$, so that both $[P_2]$ and $[P_3]$ have order 2. As $\mathbb{Z}/3\mathbb{Z}$ has no elements of order 2, $\mathrm{Cl}(F)$ can't have three elements. We conclude that $\mathrm{Cl}(F) \cong \mathbb{Z}/2\mathbb{Z}$ and $[P_2] = [P_3]$.

The identity $[P_2][P_3]^{-1} = [\mathcal{O}]$ is an example of a relation in $\mathrm{Cl}(F)$. We can prove it directly by observing that $N(3 + \sqrt{15}) = -6$. In light of (5.4.6), this means that $\mathcal{O}(3 + \sqrt{15}) = P_2 P_3$. $\qquad\square$

For larger discriminants, it's helpful to systematize the procedure for calculating the ideal class group. The following proposition is a direct combination of the two main theorems of the book—the unique factorization of ideals (Thm. 1.5.5) and the finiteness of the ideal class group (Thm. 5.1.2), along with the observation that all inert prime ideals are principal.

5.4.8 Proposition. *The ideal class group of F is generated by the finitely many classes* $[P]$, *where* P *ranges over ideals of prime norm bounded by* $NP \leq \mathfrak{M}_F$.

5.4.9 Example. Let $F = \mathbb{Q}[\sqrt{130}]$, with $\mathcal{O} = \mathbb{Z}[\sqrt{130}]$ and $D_F = 520$. By Prop. 5.4.8, a set of generators for $\mathrm{Cl}(F)$ is given by all ideals P of prime norm p satisfying

$$p \leq \frac{1}{2}\sqrt{520} \approx 11.4$$

The possible p are given in the table

p	2	3	5	7	11
$\left(\frac{130}{p}\right)$	0	1	0	1	1
b	0	1	0	2	3

Here b defines the ideal $P_p = \mathbb{Z}p + \mathbb{Z}(b + \sqrt{130})$ for which $\mathcal{O}p = P_p \bar{P}_p$.

Classes of ramified prime ideals have order at most 2: $P_2^2 = \mathcal{O} \cdot 2$, so $[P_2]^2 = [\mathcal{O}]$, and similarly $[P_5]^2 = [\mathcal{O}]$. To find other relations among the five generators, we factor a few elements of $\mathcal{O}$, the norms of which have only prime factors ≤ 11. A computer search produces the following useful elements and the corresponding relations:

$$N(12 + \sqrt{130}) = 2 \cdot 7, \qquad\qquad \mathcal{O}(12 + \sqrt{130}) = P_2 \cdot \bar{P}_7$$
$$N(35 + 3\sqrt{130}) = 5 \cdot 11, \qquad\qquad \mathcal{O}(35 + 3\sqrt{130}) = P_5 \cdot \bar{P}_{11}$$
$$N(10 + \sqrt{130}) = -2 \cdot 3 \cdot 5, \qquad\qquad \mathcal{O}(10 + \sqrt{130}) = P_2 \cdot P_3 \cdot P_5$$

From the first equation we get $[\mathcal{O}] = [P_2][\bar{P}_7] = [P_2][P_7]^{-1}$, so that $[P_7] = [P_2]$. Similarly, $[P_{11}] = [P_5]$ and $[P_3] = ([P_2][P_5])^{-1} = [P_2][P_5]$. Thus, $\mathrm{Cl}(F)$ is generated by the classes $[P_2]$ and $[P_5]$, each of order at most 2. Unless there are other relations between them, we'll have $\mathrm{Cl}(F) \cong \mathbb{Z}/2\mathbb{Z} \times \mathbb{Z}/2\mathbb{Z}$.

To prove this, it suffices to show that none of the classes $[P_2], [P_5]$ and $[P_2][P_5] = [P_3]$ equals $[\mathcal{O}]$. Lemma 5.4.7 reduces this to showing that none of the equations $x^2 - 130y^2 = \pm 2, \pm 3$ or ± 5 has integer solutions.

$x^2 - 130y^2 = \pm 5$: Reducing modulo 13, we get $x^2 = \pm 5 \pmod{13}$, which is impossible since $\left(\frac{\pm 5}{13}\right) = -1$. The same argument shows that $x^2 - 130y^2 = \pm 2$ has no solutions, so that both our generators are nontrivial.

$x^2 - 130y^2 = \pm 3$: Reducing modulo 5, this becomes $x^2 \equiv \pm 3 \pmod{5}$, which is impossible. $\qquad\square$

5.4.10 Example. Let $F = \mathbb{Q}[\sqrt{-47}]$, so that $\mathcal{O} = \mathbb{Z}[\delta]$ with $\delta^2 - \delta + 12 = 0$. Proposition 5.4.8 shows that $\mathrm{Cl}(F)$ is generated by ideals P with prime norm bounded by

$$NP \le \frac{2}{\pi}\sqrt{47} \approx 4.36.$$

We factor the two possible values of NP:

$$\mathcal{O} \cdot 2 = P_2 \bar{P}_2 \qquad\qquad P_2 = \mathbb{Z} \cdot 2 + \mathbb{Z}\delta$$
$$\mathcal{O} \cdot 3 = P_3 \bar{P}_3 \qquad\qquad P_3 = \mathbb{Z} \cdot 3 + \mathbb{Z}(-1 + \delta).$$

Any relation between $[P_2]$ and $[P_3]$ is of the form $[P_2]^a[P_3]^b = [\mathcal{O}]$. This means that $P_2^a P_3^b = \mathcal{O}\alpha$ for some $\alpha \in \mathcal{O}$ of norm $2^a 3^b$. We immediately find such an α: $N\delta = 12 = 2^2 \cdot 3$. A quick check as in Ex. 4.9.8 shows that

$$\mathcal{O}\delta = P_2^2 \bar{P}_3,$$

so that in $\mathrm{Cl}(F)$, $[P_3] = [P_2]^2$. We can replace every power of $[P_3]$ with a power of $[P_2]$, and conclude that $\mathrm{Cl}(F) = \langle [P_2] \rangle$. All that remains is to determine the order of $[P_2]$, i.e., the smallest $e > 0$ for which there is a $\beta = x + y\delta \in \mathcal{O}$ with $P_2^e = \mathcal{O}\beta$.

Taking the norm, we get

$$(5.4.11) \qquad 2^e = N(x + y\delta) = x^2 + xy + 12y^2 = \left(x + \frac{y}{2}\right)^2 + \frac{47}{4}y^2.$$

If $y = 0$, then $\beta = \pm 2^{e/2}$, so that $\mathcal{O}\beta = \mathcal{O} \cdot 2^{e/2} = P_2^{e/2}\bar{P}_2^{e/2}$. This prime factorization of $\mathcal{O}\beta$ contradicts $\mathcal{O}\beta = P_2^e$. If $y \ne 0$, we see that $N\beta \ge \lceil 47/4 \rceil = 12$, so $N\beta$ can't be 2, 4 or 8. In particular, $N\beta \ne 2$ implies that P_2 isn't principal.

If $N\beta = 16$, we must have $y = \pm 1$, otherwise the right-hand side of (5.4.11) is too big. But $x^2 + x + 12 = 16$ has no solution in $\mathbb{Z}$, which disqualifies 16 as a candidate for $N\beta$.

Next, let's try to solve $32 = x^2 + xy + 12y^2$. As before, we can't have $|y| > 1$, which leads us to a solution $(x, y) = (4, 1)$, and the factorization $\mathcal{O}(4 + \delta) = P_2^5$. Thus, $\mathrm{Cl}(F) \cong \mathbb{Z}/5\mathbb{Z}$. $\qquad\square$

The moral of the preceding example is that, for imaginary F, solving the equation $N(x + y\delta) = (x - \frac{t}{2}y)^2 - \frac{D_F}{4}y^2 = n$ is easy. Since $D_F < 0$, the norm increases with x and y, and we only need to check finitely many values of either.

5.4.12 Example. For real quadratic fields the situation is harder, as we will see in the example of $F = \mathbb{Q}[\sqrt{223}]$. We have $\mathcal{O} = \mathbb{Z}[\sqrt{223}]$ and $D_F = 892$. We start by factoring the ideals $\mathcal{O}p$ for all integer primes p bounded by

$$p \leq \frac{1}{2}\sqrt{892} \approx 14.933.$$

We make a table as in Ex. 5.4.9:

p	2	3	5	7	11	13
$\left(\frac{223}{p}\right)$	0	1	-1	-1	1	-1
b	1	1	$-$	$-$	5	$-$

The b corresponding to $p = 2, 3$ or 11 defines a prime ideal $P_p = \mathbb{Z}p + \mathbb{Z}(b + \sqrt{223})$ for which $\mathcal{O}p = P_p \bar{P}_p$. We disregard 5, 7, and 13 since they are inert, and thus belong to $[\mathcal{O}]$. We also ignore P_2, which is principal by Lemma 5.4.7, as $N(15 + \sqrt{223}) = 15^2 - 223 = 2$.

To find a relation involving $[P_3]$ and $[P_{11}]$, we look for elements of small norm divisible only by 3 and 11. Even without a computer we soon find that

$$N(16 + \sqrt{223}) = 3 \cdot 11 \quad \text{and} \quad \mathcal{O}(16 + \sqrt{223}) = P_3 \cdot P_{11}$$
$$N(14 + \sqrt{223}) = -3^3 \quad \text{and} \quad \mathcal{O}(14 + \sqrt{223}) = P_3^3$$

We conclude that $\mathrm{Cl}(F) = \langle [P_3] \rangle$ is isomorphic to $\mathbb{Z}/3\mathbb{Z}$, provided we show that $[P_3]$ is not principal. As before, this boils down to showing that the equation $x^2 - 223y^2 = \pm 3$ has no solution in $\mathbb{Z}$. We can't do this by reducing modulo a prime, because the equation $x^2 - 223y^2 \equiv \pm 3 \pmod{p}$ has a solution for each prime $p \in \mathbb{N}$. To see this, We consider several cases.

If there is an $a \in \mathbb{Z}$ with $a^2 \equiv \pm 3 \pmod{p}$, then $a^2 - 223 \cdot 0^2 \equiv \pm 3 \pmod{p}$. Assume now that both 3 and -3 are nonsquares modulo p. If 223 is a square modulo p, there exists a $b \in \mathbb{Z}$ with $b^2 \equiv 4/223 \pmod{p}$. Then $1^2 - 223b^2 \equiv -3 \pmod{p}$. Otherwise, both 3 and 223 are nonsquares, and there is a $b \in \mathbb{Z}$ satisfying $b^2 \equiv 3/223 \pmod{p}$. Then $0^2 - 223b^2 \equiv -3 \pmod{p}$.

To see that $x^2 - 223y^2 = \pm 3$ has no integral solution, we will need the techniques from the next chapter. $\square$

We extract from these examples a template for finding the ideal class group of a quadratic field F:

(a) Calculate the Minkowski bound $B = \lfloor \mathfrak{M}_F \rfloor$ as in Prop. 5.4.1, and find the prime factorizations $\mathcal{O}p = P_p \bar{P}_p$ for all noninert prime numbers $p \leq B$. The prime ideals P_p make up a list of generators for $\mathrm{Cl}(F)$, possibly highly redundant.

(b) Find a few $\alpha \in \mathcal{O}$ with $N\alpha$ small and divisible only by primes $\leq B$. Factoring $\mathcal{O}\alpha$ gives us a relation on the generators obtained in (a).

(c) Use these relations to cut down the number of generators needed, until we determine the group structure of $\mathrm{Cl}(F)$.

This template can be refined to fast algorithms, implemented in various number theory software packages. The vaguer template presented here will suffice to give you a feel for ideal class computations. Moreover, in Ch. 7 we'll learn an elementary method for computing the size of $\mathrm{Cl}(F)$, if not its full group structure.

Exercises

5.4.1. Prove Lemma 5.4.7.

5.4.2. Let I be an ideal of $\mathcal{O}$ with a norm that is divisible only by ramified primes. Prove that I is principal if and only if there exists an $\alpha \in \mathcal{O}$ with $N\alpha = \pm NI$.

5.4.3. Let $\mathcal{O}$ be the ring of integers in $\mathbb{Q}[\sqrt{D}]$. Prove that $\mathcal{O}$ is a PID when $D \in \{-1, -2, -3, -7, -11, -19, -43, -67, -163\}$. This is the complete list of imaginary quadratic $\mathcal{O}$ which are PIDs: a difficult theorem of Heegner and Stark, originally conjectured by Gauss, states that $D_F < -163$ implies $h(F) \geq 2$.

5.4.4. Prove that $h(\mathbb{Q}[\sqrt{D}]) = 1$ for $D \in \{107, 197, 227\}$.

5.4.5.* Prove that the equation $x^2 - 107y^2 = 457$ has a solution $x, y \in \mathbb{Z}$.

5.4.6.* Verify the following ideal class group calculations:

(a) $\mathrm{Cl}(\mathbb{Q}[\sqrt{10}]) \cong \mathbb{Z}/2\mathbb{Z}$
(b) $\mathrm{Cl}(\mathbb{Q}[\sqrt{-21}]) \cong \mathbb{Z}/2\mathbb{Z} \times \mathbb{Z}/2\mathbb{Z}$
(c) $\mathrm{Cl}(\mathbb{Q}[\sqrt{399}]) \cong \mathbb{Z}/4\mathbb{Z} \times \mathbb{Z}/2\mathbb{Z}$ or $\mathbb{Z}/2\mathbb{Z} \times \mathbb{Z}/2\mathbb{Z}$ (to decide which, we'll need the techniques of Ch. 6)
(d) $\mathrm{Cl}(\mathbb{Q}[\sqrt{-71}]) \cong \mathbb{Z}/7\mathbb{Z}$
(e) $\mathrm{Cl}(\mathbb{Q}[\sqrt{-23}]) \cong \mathbb{Z}/3\mathbb{Z}$

5.4.7. Compute $\mathrm{Cl}(\mathbb{Q}[\sqrt{D}])$ for $D \in \{-38, -26, 79, 195\}$.

5.4.8.* Given any $n \geq 3$, find a $D < 0$ for which $\mathrm{Cl}(\mathbb{Q}[\sqrt{D}])$ contains an ideal class of order n.

5.4.9.* Let $F = \mathbb{Q}[\sqrt{D}]$ be an imaginary quadratic field, with D square-free. Let $p_1, \ldots, p_s$ be the prime factors of D, and $p_1, \ldots, p_r$ the prime factors of D_F (the ramified primes). Since $D_F = D$ or $4D$, we have that $r = s$ unless $D \equiv 3 \pmod 4$, when $r = s + 1$ and $p_r = 2$. Define prime ideals P_i by $\mathcal{O}p_i = P_i^2$ for $1 \leq i \leq r$. Denote by $\mathrm{Cl}(F)[2]$ the group of ideal classes of order dividing 2.

(a) Show that each element of $\mathrm{Cl}(F)[2]$ is a product of distinct classes chosen from $[P_1], \ldots, [P_r]$.

(b) Factor $\mathcal{O}\sqrt{D}$ to find a relation among $[P_1], \ldots, [P_s]$.
(c) Use Exer. 5.4.2 to show that we have a relation $[P_{i_1}] \cdots [P_{i_k}] = [\mathcal{O}]$, $1 \le i_j \le r$ if and only if there is an $\alpha \in \mathcal{O}$ with $N\alpha = p_{i_1} \cdots p_{i_k}$.
(d) Prove that $N\alpha = p_{i_1} \ldots p_{i_k}$ has a solution if and only if $\{i_1, \ldots, i_k\} = \{1, \ldots, s\}$, so that the relation we found in (a) is the only one.

5.4.10. Give a necessary and sufficient condition on $D < 0$ for $h(\mathbb{Q}[\sqrt{D}])$ to be odd.

Chapter 6
Continued Fractions

6.1 Motivation

When we write $\pi = 3.141592\ldots$, we really mean that π be approximated (the "$\ldots$" part) by the rational number $3.141592 = \frac{3141592}{1000000}$. That approximation requires a large numerator and denominator to accurately give the first six decimal places of π. We can attain the same precision with a shorter fraction

$$(6.1.1) \qquad \frac{355}{113} = 3.14159292035\ldots.$$

The Chinese mathematician Zu Chongzhi discovered this "detailed approximation" in the fifth century CE. With an error of just 0.0000085%, it was the best approximation of π known for the subsequent millennium.

The theory of continued fractions gives a method for efficiently approximating real numbers by rationals, as in (6.1.1). A crude approximation to $\eta \in \mathbb{R}$ is given by the largest integer not greater than η. It is called the **floor** of η, denoted $\lfloor \eta \rfloor$, and characterized by

$$(6.1.2) \qquad \lfloor \eta \rfloor \in \mathbb{Z}, \quad 0 \le \eta - \lfloor \eta \rfloor < 1.$$

To estimate the error $\eta - \lfloor \eta \rfloor$, we invert it and apply the floor function again. Repeating this gives ever closer rational approximations to η.

6.1.3 Example. Here's an illustration of the procedure:

$$\pi = 3 + \cfrac{1}{\cfrac{1}{0.141\ldots}} = 3 + \cfrac{1}{7.062\ldots} = 3 + \cfrac{1}{7 + \cfrac{1}{\cfrac{1}{0.062\ldots}}} = 3 + \cfrac{1}{7 + \cfrac{1}{15.997\ldots}}$$

$$(6.1.4)$$

$$= 3 + \cfrac{1}{7 + \cfrac{1}{15 + \cfrac{1}{\cfrac{1}{0.997\ldots}}}} = 3 + \cfrac{1}{7 + \cfrac{1}{15 + \cfrac{1}{1.003\ldots}}} = \ldots$$

M. Trifković, *Algebraic Theory of Quadratic Numbers*, Universitext,
DOI 10.1007/978-1-4614-7717-4_6, © Springer Science+Business Media New York 2013

Rounding off the last denominator to 1 doesn't change the value of the expression too much, and yields the approximation (6.1.1):

$$\pi \approx 3 + \cfrac{1}{7 + \cfrac{1}{15 + \frac{1}{1}}} = \frac{355}{113}$$

The sequence of integers $3, 7, 15, 1, \ldots$ is called the CF (for "Continued Fraction") sequence associated with π. □

6.1.5 Definition. *A* **CF sequence** *is a finite or infinite sequence* $a_0, a_1, \ldots$ *satisfying* $a_i \in \mathbb{Z}$ *for* $i \geq 0$ *and* $a_i > 0$ *for* $i \geq 1$. *The* a_i *are called the* **elements** *of the CF sequence.*

The computation in Ex. 6.1.3 is an instance of the general recipe for assigning a CF sequence to a real number.

THE CONTINUED FRACTION PROCEDURE (CFP)

1. INPUT: $\eta \in \mathbb{R}$. Put $\eta_0 := \eta$ and set counter $i := 0$.
2. Put $a_i := \lfloor \eta_i \rfloor$. Print a_i, and continue.
3. If $a_i = \eta_i$, STOP.
4. Otherwise, set $\eta_{i+1} := 1/(\eta_i - a_i)$, increase the counter $i := i + 1$, and go to Step 2.

It's not hard to see that the a_i produced by the Continued Fraction Procedure indeed form a CF sequence. The η_i is called the ith **tail** of η, for reasons explained in Exer. 6.1.8.

Technically, these steps do not constitute an algorithm: if η is irrational, we never get to Step 3 and are therefore caught in an infinite loop. In practice, we stop once we have enough a_i's to approximate η with the desired precision. We are especially interested in applying the Continued Fraction Procedure (CFP) to the sort of numbers we've studied.

6.1.6 Definition. *A* **quadratic number** *is an irrational element of a quadratic field. In other words, it's a complex number of the form*

$$\frac{x + y\sqrt{z}}{w},$$

with $x, y, z, w \in \mathbb{Z}, y, w \neq 0$, *and* z *not a perfect square.*

Note that, unlike in the previous chapters, we require quadratic numbers to be irrational. We say that a quadratic number is imaginary if $z < 0$, and real if $z > 0$. In this chapter, we deal exclusively with quadratic numbers in $\mathbb{R}$, so we drop the qualifier "real." We also agree that $\sqrt{z} > 0$.

In previous chapters we worked inside a fixed quadratic field. Now we shift our perspective, and aim to distinguish the set of all quadratic numbers from other real numbers. The main result of this chapter is that they are precisely the real numbers whose continued fractions eventually become periodic. This will complete our understanding of quadratic fields: the description and explicit computation of units in a real quadratic field.

Exercises

6.1.1. Use the CFP to find the first five elements of the continued fraction of: (a) $-249/41$; (b) $(1 + \sqrt{5})/2$; (c) $\sqrt[3]{10}$.

6.1.2. Find the first 15 elements of the continued fraction of e. Do you see a pattern? Can you prove it? The pattern is a little easier to state if you consider $e - 1$ instead.

6.1.3. Take $\eta = \pi$ and compute η_i for $i = 0, 1, 2, 3$. Compare with the calculation in (6.1.4).

6.1.4. Arrange the numbers

$$\pi, \quad 3, \quad \frac{22}{7}, \quad \frac{333}{106}, \quad \frac{355}{113}, \quad \frac{103993}{33102}, \quad \frac{104348}{33215}$$

in increasing order. What do you observe? Where do these fractions come from? (Archimedes knew that $22/7$ is a good approximation of π. More impressively, he knew it wasn't the exact value, but contented himself with asserting that $3 < \pi < 22/7$.)

6.1.5. Prove that the a_i produced by the CFP form a CF sequence.

6.1.6. Let $a, b \in \mathbb{Z}, b > 0$. The division algorithm produces $q, r \in \mathbb{Z}$ such that $a = qb + r$ with $0 \le r < b$. Fill each blank in

$$\frac{a}{b} = ? + \frac{1}{\frac{?}{?}}$$

with one of the letters a, b, q, r to get the result of the first iteration of the CFP on $\eta = a/b$. What is the connection between the division algorithm and the CFP applied to $\eta \in \mathbb{Q}$?

6.1.7. Show that $\eta \in \mathbb{Q}$ if and only if the CFP applied to η terminates after producing a finite sequence $a_0, \ldots, a_k$. In that case, observe that

$$\eta = a_0 + \cfrac{1}{a_1 + \cfrac{1}{\ddots a_{k-1} + \frac{1}{a_k}}}$$

6.1.8. Prove the following identity, which justifies calling η_i the ith tail of η:

$$\eta = a_0 + \cfrac{1}{a_1 + \cfrac{1}{\ddots a_{i-1} + \frac{1}{\eta_i}}}.$$

6.1.9. Prove that there exists a $\left[\begin{smallmatrix} a & b \\ c & d \end{smallmatrix}\right] \in \mathrm{GL}_2(\mathbb{Z})$ such that

$$\eta_i = \frac{a\eta + b}{c\eta + d}.$$

Conclude that η is irrational if and only if all its tails η_i are irrational.

6.1.10. Prove that η is a quadratic number if and only if it satisfies a quadratic equation $a\eta^2 + b\eta + c = 0$ with coefficients in $\mathbb{Z}$ and a non-square discriminant $b^2 - 4ac$.

6.2 Finite and Infinite Continued Fractions

The inverse to the Continued Fraction Procedure should send an arbitrary CF sequence to a real number. We define it first for *finite* CF sequences. For technical convenience, in this section we drop the integrality requirement, and consider a sequence $\{a_i\}$ of *real* numbers with $a_i > 0$ for $i \geq 1$.

6.2.1 Definition. *The* **finite continued fraction** *with elements* $a_0, \ldots, a_k$ *is the expression*

$$[a_0, a_1, a_2, \ldots, a_k] = a_0 + \cfrac{1}{a_1 + \cfrac{1}{a_2 + \cfrac{1}{\ddots\, a_{k-1} + \cfrac{1}{a_k}}}}.$$

Since $a_i > 0$ for $i \geq 1$, we never divide by zero when computing $[a_0, \ldots, a_k]$. As you showed in Exer. 6.1.6, every rational number is equal to a finite continued fraction with elements in $\mathbb{Z}$.

6.2.2 Example. To find the continued fraction for $57/17$, we repeatedly apply the division algorithm:

$$\frac{57}{17} = \frac{3 \cdot 17 + 6}{17} = 3 + \frac{6}{17} = 3 + \frac{1}{\frac{17}{6}} = 3 + \cfrac{1}{2 + \frac{5}{6}} = 3 + \cfrac{1}{2 + \cfrac{1}{1 + \frac{1}{5}}}.$$

In continued fraction notation, $57/17 = [3, 2, 1, 5]$. □

6.2.3 Definition. *The* ith **convergent** *of the sequence* $\{a_i\}$ *is the truncated continued fraction* $[a_0, a_1, \ldots, a_i]$.

6.2.4 Example. Other than $57/17$, the convergents of $[3, 2, 1, 5]$ are

$$[3] = 3, \qquad [3,2] = 3 + \frac{1}{2} = \frac{7}{2}, \qquad [3,2,1] = 3 + \cfrac{1}{2 + \cfrac{1}{1}} = \frac{10}{3}. \qquad \Box$$

Since the a_i are computed using the recursive CFP, it isn't surprising that there is a recursive formula for the numerator and denominator of a convergent.

6.2.5 Proposition. *Define the sequences* $\{p_i\}, \{q_i\}$ *for* $-2 \le i$:

(6.2.6)
$$\begin{aligned} p_{-2} = 0, \; p_{-1} = 1, \quad p_{i+1} = a_{i+1}p_i + p_{i-1}, \\ q_{-2} = 1, \; q_{-1} = 0, \quad q_{i+1} = a_{i+1}q_i + q_{i-1}. \end{aligned}$$

Then $[a_0, \ldots, a_i] = p_i/q_i$.

Unlike in previous chapters, there is no suggestion that p_i is prime. The values of p_i and q_i depend only on $a_0, \ldots, a_i$, since no a_j with $j > i$ enters into computing them. When we want to emphasize that fact, we write $p_i(a_0, \ldots, a_i)$ instead of p_i.

Proof. We argue by induction. The base case is clear, as $p_0 = a_0$ and $q_0 = 1$. For the induction step, assume that the recursions are valid up to the jth convergent of *any* sequence a_i satisfying $a_i \in \mathbb{R}, a_i > 0$ for $i > 1$, in particular of the right side of the equality

$$[a_0, a_1, \ldots, a_{j-1}, a_j, a_{j+1}] = \left[a_0, a_1, \ldots, a_{j-1}, a_j + \frac{1}{a_{j+1}} \right].$$

Let p_i, q_i be the numbers defined by (6.2.6) for the left side, and p_i', q_i' the corresponding quantities for the right side. Since the elements of the two sides with indices from 0 to $j-1$ coincide, the corresponding convergents are the same: $p_i = p_i', q_i = q_i'$ for all $0 \le i \le j-1$. Thus,

$$[a_0, \ldots, a_{j+1}] = \left[a_0, \ldots, a_j + \frac{1}{a_{j+1}} \right] = \frac{\left(a_j + \frac{1}{a_{j+1}} \right) p_{j-1}' + p_{j-2}'}{\left(a_j + \frac{1}{a_{j+1}} \right) q_{j-1}' + q_{j-2}'}$$

$$= \frac{\left(a_j + \frac{1}{a_{j+1}} \right) p_{j-1} + p_{j-2}}{\left(a_j + \frac{1}{a_{j+1}} \right) q_{j-1} + q_{j-2}} = \frac{a_{j+1}(a_j p_{j-1} + p_{j-2}) + p_{j-1}}{a_{j+1}(a_j q_{j-1} + q_{j-2}) + q_{j-1}}$$

$$= \frac{a_{j+1}p_j + p_{j-1}}{a_{j+1}q_j + q_{j+1}} = \frac{p_{j+1}}{q_{j+1}}. \qquad \blacksquare$$

6.2.7 Example. We use the CF sequence of π, the first four terms of which we computed in Ex. 6.1.3 to illustrate a convenient way of tabulating the convergents of a continued fraction:

i	-2	-1	0	1	2	3	4	5	...
a_i	–	–	3	7	15	1	292	1	...
p_i	0	1	3	22	333	355	103993	104348	...
q_i	1	0	1	7	106	113	33102	33215	...

We start by putting $\left[\begin{smallmatrix}0\\1\end{smallmatrix}\right]$ and $\left[\begin{smallmatrix}1\\0\end{smallmatrix}\right]$ in columns numbered -2 and -1, and then write the CF sequence across the top. An example of the pattern by which we fill either row is:

$$
\begin{array}{c}
15 \\
\nwarrow \quad \uparrow = \\
3 \xleftarrow{\;+\;} 22 \quad 333
\end{array}
\qquad 333 = 15 \cdot 22 + 3.
$$

Reading along the arrows gives the recursion (6.2.6). This recipe computes both the p_i and the q_i, since their recursions differ only in the initial conditions. □

6.2.8 Corollary. *The convergents p_i/q_i satisfy the following:*

(a) *The recursions (6.2.6) can be written in matrix form:*

$$
M_{i+1} = M_i A_{i+1}, \quad \text{where } M_{-1} = \left[\begin{smallmatrix}0 & 1\\1 & 0\end{smallmatrix}\right],\, M_i = \left[\begin{smallmatrix}p_{i-1} & p_i\\q_{i-1} & q_i\end{smallmatrix}\right], \text{ and } A_i = \left[\begin{smallmatrix}0 & 1\\1 & a_i\end{smallmatrix}\right].
$$

(b) *We have $p_{i-1}q_i - p_iq_{i-1} = (-1)^i$. In particular, when $\{a_i\}$ is a CF sequence ($a_i \in \mathbb{Z}$), we have $\gcd(p_i, q_i) = 1$ and the fraction p_i/q_i is in lowest terms.*

(c) *The ordering of the convergents is given by*

$$
\frac{p_0}{q_0} < \frac{p_2}{q_2} < \frac{p_4}{q_4} < \cdots < \frac{p_5}{q_5} < \frac{p_3}{q_3} < \frac{p_1}{q_1}.
$$

Proof.

(a) Multiply the matrices.

(b) Taking the determinant of $M_{i+1} = M_i A_{i+1}$ yields $p_i q_{i+1} - p_{i+1}q_i = -(p_{i-1}q_i - p_iq_{i-1})$, so that (b) follows by induction from the base case $\det M_{-1} = -1$.

(c) Using the recurrence relation (6.2.6) we compute

$$
\frac{p_{i+2}}{q_{i+2}} - \frac{p_i}{q_i} = \frac{(-1)^i a_{i+2}}{q_i q_{i+2}}.
$$

Since $a_i > 0$ for $i \geq 1$, we have $p_0/q_0 < p_2/q_2 < \cdots$ and $\cdots < p_3/q_3 < p_1/q_1$. To put the two sequences of inequalities together, observe that

$$
\frac{p_{i+1}}{q_{i+1}} - \frac{p_i}{q_i} = \frac{p_{i+1}q_i - p_iq_{i+1}}{q_{i+1}q_i} = \frac{(-1)^i}{q_{i+1}q_i},
$$

so that $p_{2i}/q_{2i} < p_{2i+1}/q_{2i+1} < p_{2j+1}/q_{2j+1}$ for any $i > j \geq 0$. ∎

We now turn to defining infinite continued fractions. We've seen in Ex. 6.1.3 and Exer. 6.1.4 that the convergents $[3, 7, 15, 1, \ldots]$ give closer and closer approximations of π. This suggests defining an infinite continued fraction as the limit of its finite convergents.

6.2.9 Proposition-Definition. *When* $a_0, a_1, a_2 \ldots$ *is an infinite CF sequence, the limit*

$$\lim_{i \to \infty} [a_0, a_1, \ldots, a_i]$$

exists and is irrational. We denote it $[a_0, a_1, \ldots]$, *and call it the* **infinite continued fraction** *with elements* $a_0, a_1, \ldots$.

Proof. By Cor. 6.2.8 (c), there is a sequence of nested intervals

$$\left[\frac{p_0}{q_0}, \frac{p_1}{q_1} \right] \supset \left[\frac{p_2}{q_2}, \frac{p_3}{q_3} \right] \supset \left[\frac{p_4}{q_4}, \frac{p_5}{q_5} \right] \supset \cdots$$

Let η belong to their intersection, which is non-empty by completeness of $\mathbb{R}$. Comparing the distances from p_i/q_i to η and p_{i+1}/q_{i+1}, we get the estimate

$$\left| \eta - \frac{p_i}{q_i} \right| < \left| \frac{p_{i+1}}{q_{i+1}} - \frac{p_i}{q_i} \right| = \left| \frac{(-1)^{i+1}}{q_i q_{i+1}} \right| = \frac{1}{q_i q_{i+1}}.$$

As $\{q_i\}_{i \geq 0}$ is an increasing sequence of natural numbers, we conclude that $\eta = \lim_{i \to \infty} p_i/q_i$. In particular, $\eta \neq p_i/q_i$ for all i.

If $\eta = a/b$ with $b > 0$, then $a/b \neq p_i/q_i$ implies $|a q_i - b p_i| \geq 1$. This gives, for all i:

$$\frac{1}{b q_i} \leq \left| \frac{a}{b} - \frac{p_i}{q_i} \right| < \frac{1}{q_i q_{i+1}}.$$

Thus, $q_{i+1} < b$ for all i, a contradiction that shows η to be irrational. ∎

6.2.10 Definition. *A* **continued fraction of** $\eta \in \mathbb{R}$ *is an equality of the form* $\eta = [a_0, a_1, \ldots]$, *where* $a_0, a_1, \ldots$ *is a CF sequence, finite if and only if* $\eta \in \mathbb{Q}$.

Exercises

6.2.1. Prove that $q \in \mathbb{Q}$ has precisely two continued fractions, namely $q = [a_0, \ldots, a_k] = [a_0, \ldots, a_k - 1, 1]$, where $a_k \neq 1$.

6.2.2. Prove the following about the convergents of a continued fraction $[a_0, a_1, \ldots]$:

(a) $q_{i+1} > q_i > 0$ for $i \geq 0$
(b) $p_{i+2} q_i - p_i q_{i+2} = (-1)^i a_{i+2}$

6.2.3. Let $\eta = [a_0, a_1, \ldots] \in \mathbb{R}$, with convergents p_i/q_i. For $a \in \mathbb{Z}$, find the continued fraction for $\eta' = \eta + a$. Prove that its convergents p_i'/q_i' satisfy $p_i' = p_i + a q_i, q_i' = q_i$, for all $i \geq 0$.

6.2.4. Recall that we write $p_i(a_0, \ldots, a_i), q_i(a_0, \ldots, a_i)$ when we want to emphasize that p_i and q_i depend only on $a_0, \ldots, a_i$. Prove that:

(a) $q_i(a_0, \ldots, a_i) = p_{i-1}(a_1, \ldots, a_i)$
(b) $p_i(a_0, \ldots, a_i) = p_i(a_i, \ldots, a_0)$
(c) $q_{i-1}(a_0, \ldots, a_{i-1})/p_{i-1}(a_0, \ldots, a_{i-1}) = [a_0, \ldots, a_i]^{-1}$.

6.2.5. Give an algorithm for deciding which of the (finite or infinite) continued fractions $[a_0, a_1, \ldots]$ and $[b_0, b_1, \ldots]$ is bigger.

6.2.6. Give a necessary and sufficient condition on η for its continued fraction to satisfy $a_1 = 1$.

6.2.7. Let $\eta = [a_0, a_1, \ldots]$, finite or infinite. Find the continued fraction for $-\eta$.

6.3 Continued Fraction of an Irrational Number

In this section we establish a one-to-one correspondence between irrational numbers and CF sequences,

$$\mathbb{R} \setminus \mathbb{Q} \underset{\text{LP}}{\overset{\text{CFP}}{\rightleftarrows}} \{(a_i)_{i \geq 0} : a_i \in \mathbb{Z}, a_i > 0 \text{ for } i \geq 1\}.$$

The function LP is the Limit Procedure of Prop.-Def. 6.2.9. It associates to a CF sequence $a_0, a_1, \ldots$ the real number

$$[a_0, a_1, \ldots] = \lim_{i \to \infty} [a_0, \ldots, a_i].$$

The function CFP is the Continued Fraction Procedure, which inputs an irrational number η and outputs the CF sequence $a_0, a_1, \ldots$ defined by the recursions

$$\eta_0 = \eta, \quad a_i = \lfloor \eta_i \rfloor, \quad \eta_{i+1} = \frac{1}{\eta_i - a_i}.$$

Since the η_i are all irrational, we never come to Step 3 of the CFP, so $\eta_i - a_i$ is never 0. As usual, put $p_i/q_i = [a_0, \ldots, a_i]$.

6.3.1 Lemma. *With notations as above,* $\eta_{i+1} = \frac{-q_{i-1}\eta + p_{i-1}}{q_i \eta - p_i}$.

Proof. By the recursions for convergents we have:

(6.3.2) $$\eta = [a_0, a_1, \ldots, a_i, \eta_{i+1}] = \frac{\eta_{i+1} p_i + p_{i-1}}{\eta_{i+1} q_i + q_{i-1}}.$$

The last equality comes from the recursions (6.2.6), with $a_{i+1} = \eta_{i+1}$. Solving for η_{i+1} proves the lemma. ∎

We need to estimate how closely η is approximated by its convergents.

6.3.3 Lemma. *For $\eta \in \mathbb{R}\backslash\mathbb{Q}$ we have the error estimate*

$$\frac{1}{q_i(q_i + q_{i+1})} < \left|\eta - \frac{p_i}{q_i}\right| < \frac{1}{q_i q_{i+1}}.$$

Proof. The expression (6.3.2) yields an exact formula for the error of approximating η by p_i/q_i:

$$\left|\eta - \frac{p_i}{q_i}\right| = \left|\frac{\eta_{i+1}p_i + p_{i-1}}{\eta_{i+1}q_i + q_{i-1}} - \frac{p_i}{q_i}\right| = \left|\frac{p_{i-1}q_i - p_i q_{i-1}}{q_i(\eta_{i+1}q_i + q_{i-1})}\right|$$

$$= \frac{1}{q_i(\eta_{i+1}q_i + q_{i-1})}.$$

We convert this into a more usable estimate by using $a_{i+1} < \eta_{i+1} < a_{i+1} + 1$:

$$q_{i+1} = a_{i+1}q_i + q_{i-1} < \eta_{i+1}q_i + q_{i-1} < (a_{i+1} + 1)q_i + q_{i-1} = q_{i+1} + q_i.$$

Combining the two identities proves the lemma. ∎

6.3.4 Proposition. *The functions CFP and LP are mutually inverse.*

Proof. We only show that $\text{LP} \circ \text{CFP} = \text{id}$, and leave the other composition as an exercise. Start with an irrational η and define a_i, p_i, q_i as usual. Passing to the limit as $i \to \infty$ in

$$\left|\eta - \frac{p_i}{q_i}\right| < \frac{1}{q_i q_{i+1}} < \frac{1}{q_i^2},$$

we get that $\eta = \lim_{i \to \infty}(p_i/q_i) = \lim_{i \to \infty}[a_0, \ldots, a_i] = [a_0, a_1, \ldots]$, as desired. ∎

Exercises

6.3.1. Show that $\text{CFP} \circ \text{LP} = \text{id}$. In other words, start with an infinite CF sequence $a_0, a_1, \ldots$ and put $\eta = \lim_{i \to \infty}[a_0, \ldots, a_i]$. Prove that applying the CFP to η recovers the sequence $a_0, a_1, \ldots$.

6.3.2. Show that $|\eta - p_{i+1}/q_{i+1}| < |\eta - p_i/q_i|$.

6.3.3.* Arrange the numbers

$$\eta, \quad \frac{p_i}{q_i}, \quad \frac{p_{i+1}}{q_{i+1}}, \quad \frac{p_i + p_{i+1}}{q_i + q_{i+1}}$$

in increasing order. Assume that i is even; the case of i odd is similar.

6.3.4. We say that $\eta, \eta' \in \mathbb{R} \setminus \mathbb{Q}$ **have the same tail** if and only if $\eta_i = \eta'_j$ for some $i, j \geq 0$. If this happens, show that $\eta' = (a\eta + b)/(c\eta + d)$ for some $\left[\begin{smallmatrix} a & b \\ c & d \end{smallmatrix}\right] \in \mathrm{GL}_2(\mathbb{Z})$. If, moreover, $i \equiv j \pmod{2}$, prove that we may take $\left[\begin{smallmatrix} a & b \\ c & d \end{smallmatrix}\right] \in \mathrm{SL}_2(\mathbb{Z})$.

6.3.5. Use the bounds of Lemma 6.3.3 to find the irrational number with slowest-converging convergents.

6.4 Periodic Continued Fractions

We now describe the continued fraction of a quadratic number. The results in this section are not only beautiful, but also essential to finding units in real quadratic fields, and to classifying quadratic forms in Ch. 7.

6.4.1 Definition. *An infinite continued fraction* $[a_0, a_1, a_2, \ldots]$ *is* **periodic** *if there exist* $l \in \mathbb{N}, r \in \mathbb{Z}_{\geq 0}$ *such that* $a_k = a_{k+l}$ *for all* $k \geq r$. *If we can take* $r = 0$, *we say that the continued fraction is* **purely periodic**.

A periodic continued fraction looks like

$$[b_0, \ldots, b_{r-1}, c_0, \ldots, c_{l-1}, c_0, \ldots, c_{l-1}, \ldots].$$

The shortest repeating part $c_0, \ldots, c_{l-1}$ is called the **period**, and the corresponding l the **period length**. We abbreviate a periodic continued fraction by putting a bar over the period: $[b_0, \ldots, b_{r-1}, \overline{c_0, \ldots, c_{l-1}}]$.

6.4.2 Theorem. *The continued fraction of* $\eta \in \mathbb{R} \setminus \mathbb{Q}$ *is periodic if and only if* η *is a quadratic number.*

We will prove each implication as a separate proposition preceded by an example. The numbers η_i, p_i, q_i, etc., are defined as usual in reference to η.

6.4.3 Example. Let's compute the value of the periodic continued fraction $\eta = [3, 1, \overline{2, 5}] = [3, 1, 2, 5, 2, 5 \ldots]$. The purely periodic part $\eta_2 = [2, 5, 2, 5, \ldots]$ doesn't change when we remove the first two elements:

$$\eta_2 = [2, 5, 2, 5, \ldots] = [2, 5, \eta_2] = 2 + \cfrac{1}{5 + \cfrac{1}{\eta_2}} = \frac{11\eta_2 + 2}{5\eta_2 + 1}.$$

Clearing denominators we get a quadratic equation $5\eta_2^2 - 10\eta_2 - 2 = 0$ with solutions $(5 \pm \sqrt{35})/5$. Since $\lfloor \eta_2 \rfloor = 2$, we must have $\eta_2 = (5 + \sqrt{35})/5$. We then compute:

$$\eta = [3, 1, \eta_2] = 3 + \cfrac{1}{1 + \cfrac{1}{\eta_2}} = 3 + \cfrac{1}{1 + \cfrac{5}{5 + \sqrt{35}}} = \frac{42 + \sqrt{35}}{13}.$$

$\square$

6.4.4 Proposition. *If the continued fraction of η is periodic, then η is a quadratic number.*

Proof. The proof is an immediate generalization of the preceding computation. We consider two cases:

(a) If $\eta = [\overline{c_0, \ldots, c_{l-1}}]$ is purely periodic, then

$$\eta = [c_0, \ldots, c_{l-1}, \eta] = \frac{\eta p_{l-1} + p_{l-2}}{\eta q_{l-1} + q_{l-2}}.$$

This yields a quadratic equation satisfied by η:

$$q_{l-1}\eta^2 + (q_{l-2} - p_{l-1})\eta - p_{l-2} = 0.$$

(b) In general, let $\eta = [b_0, \ldots, b_{r-1}, \zeta]$, where ζ is the purely periodic part. Putting $[b_0, \ldots, b_i] = h_i/k_i$, we find

$$\eta = \frac{\zeta h_{r-1} + h_{r-2}}{\zeta k_{r-1} + k_{r-2}} \in \mathbb{Q}[\zeta],$$

because $\mathbb{Q}[\zeta]$ is a quadratic field by (a). Thus η is also a quadratic number. ∎

6.4.5 Example. As a warm-up for proving that the continued fraction of a quadratic number is periodic, consider the following application of the CFP:

$$\eta = \frac{6+\sqrt{10}}{2} = 4 + \frac{\sqrt{10}-2}{2} = 4 + \cfrac{1}{\frac{2}{\sqrt{10}-2}} = 4 + \cfrac{1}{\frac{2+\sqrt{10}}{3}}$$

$$= 4 + \cfrac{1}{1 + \frac{\sqrt{10}-1}{3}} = 4 + \cfrac{1}{1 + \cfrac{1}{\frac{3}{\sqrt{10}-1}}} = 4 + \cfrac{1}{1 + \cfrac{1}{\frac{1+\sqrt{10}}{3}}}$$

$$= 4 + \cfrac{1}{1 + \cfrac{1}{1 + \frac{\sqrt{10}-2}{3}}} = 4 + \cfrac{1}{1 + \cfrac{1}{1 + \cfrac{1}{\frac{3}{\sqrt{10}-2}}}} = 4 + \cfrac{1}{1 + \cfrac{1}{1 + \cfrac{1}{\frac{2+\sqrt{10}}{2}}}}$$

$$= 4 + \cfrac{1}{1 + \cfrac{1}{1 + \cfrac{1}{2 + \frac{\sqrt{10}-2}{2}}}} = 4 + \cfrac{1}{1 + \cfrac{1}{1 + \cfrac{1}{2 + \cfrac{1}{\frac{2}{\sqrt{10}-2}}}}} = 4 + \cfrac{1}{1 + \cfrac{1}{1 + \cfrac{1}{2 + \cfrac{1}{\frac{2+\sqrt{10}}{3}}}}}$$

Each line consists of three expressions, together constituting a single iteration of the CFP. All action takes place in the denominator of the last $1/\ldots$ fraction in the preceding line. In the first expression we write $\eta_i = a_i + (\eta_i - a_i)$, while in the second we invert the fractional part to get

$$\eta_i = a_i + \cfrac{1}{\frac{1}{\eta_i - a_i}} = a_i + \frac{1}{\eta_{i+1}}.$$

The third expression is peculiar to quadratic numbers: we rationalize the denominator of η_{i+1}. We read off η_i as the denominator of the last $1/\ldots$ fraction on the ith line:

$$\eta_1 = \frac{2+\sqrt{10}}{3}, \ \eta_2 = \frac{1+\sqrt{10}}{3}, \ \eta_3 = \frac{2+\sqrt{10}}{2}, \ \eta_4 = \frac{2+\sqrt{10}}{3}.$$

Since $\eta_4 = \eta_1$, and since a_i and η_{i+1} depend only on η_i, we see that $\eta_5 = \eta_2$. In general, $\eta_{i+3} = \eta_i$ and $a_{i+3} = a_i$ for $i \geq 1$, hence $(6+\sqrt{10})/2 = [4,\overline{1,1,2}]$. $\qquad\square$

All η_i in the example have the form $\eta_i = (m_i + \sqrt{10})/v_i$ for $m_i, v_i \in \mathbb{Z}$. The appearance of $\sqrt{10}$ is somewhat arbitrary. Other elements of $\mathbb{Q}[\sqrt{10}]$ can be used instead, most naturally η itself. Indeed, $\sqrt{10} = 2\eta - 6$, so that

$$\eta_1 = \frac{-4+2\eta}{3}, \ \eta_2 = \frac{-5+2\eta}{3}, \ \eta_3 = \frac{-4+2\eta}{2}, \ \eta_4 = \frac{-4+2\eta}{3}.$$

Note that 2η is the smallest multiple of η that is in $\mathcal{O}$.

In general, a quadratic number η satisfies $e\eta^2 - f\eta + g = 0$ with $e, f, g \in \mathbb{Z}$. Put $\eta_{\mathcal{O}} = e\eta$ or $e\bar{\eta}$, whichever is bigger. Then $\eta_{\mathcal{O}}$ is a root of the polynomial $P(x) = x^2 - fx + eg$. In particular, $\eta_{\mathcal{O}}$ is an algebraic integer.

6.4.6 Proposition. *If $\eta \in \mathbb{R}$ is a quadratic number, then the continued fraction of η is periodic.*

Proof. By constructing an explicit recursion, we will first find $m_i, v_i \in \mathbb{Z}$ satisfying

$$(6.4.7) \qquad\qquad \eta_i = \frac{m_i + \eta_{\mathcal{O}}}{v_i}, \text{ for } i \geq 0.$$

We will then prove that the m_i and v_i are bounded independently of i. Since they are in $\mathbb{Z}$, there are only finitely many different pairs (m_i, v_i). For some $i \neq j$ we must have $m_i = m_j$ and $v_i = v_j$, hence also $\eta_i = \eta_j$. The periodicity of the a_i will then follow as in Ex. 6.4.5.

Recursion for (m_i, v_i): We substitute (6.4.7) into the formula $\eta_{i+1} = 1/(\eta_i - a_i)$:

$$\eta_{i+1} = \frac{1}{\frac{m_i+\eta_{\mathcal{O}}}{v_i} - a_i} = \frac{v_i}{(m_i - a_i v_i) + \eta_{\mathcal{O}}} \cdot \frac{m_i - a_i v_i + \bar{\eta}_{\mathcal{O}}}{m_i - a_i v_i + \bar{\eta}_{\mathcal{O}}}$$

$$= \frac{v_i(m_i - a_i v_i + \bar{\eta}_{\mathcal{O}})}{P(a_i v_i - m_i)} = \frac{a_i v_i - m_i - f + \eta_{\mathcal{O}}}{-\frac{P(a_i v_i - m_i)}{v_i}}.$$

This suggests putting $m_{i+1} = a_i v_i - m_i - f$ and $v_{i+1} = -P(a_i v_i - m_i)/v_i = -P(m_{i+1} + f)/v_i$. Using the identity $P(x + f) = P(-x)$, we get the final form of the desired recursions:

$$(6.4.8) \qquad\qquad m_{i+1} = a_i v_i - m_i - f, \quad v_{i+1} = -\frac{P(-m_{i+1})}{v_i}.$$

Since the roots of $P(x)$ are irrational, $v_i v_{i-1} = P(-m_i) \neq 0$. By induction, $v_i \neq 0$ and v_{i+1} is well-defined. Finally, we need the initial condition:

$$\text{if } \eta_O = e\eta, \qquad\qquad \text{take } m_0 = 0, \; v_0 = e;$$
$$\text{if } \eta_O = e\bar{\eta} = f - e\eta, \quad \text{take } m_0 = -f, \; v_0 = -e.$$

In either case, $\eta = (m_0 + \eta_O)/v_0$ and $P(-m_0) \equiv 0 \pmod{v_0}$.

Integrality of m_i and v_i: The preceding identities are the case $i = 0$ of the statement

$$m_i, v_i \in \mathbb{Z} \text{ and } P(-m_i) \equiv 0 \pmod{v_i},$$

which we prove by induction. It is clear from the recursion (6.4.8) that $m_{i+1} \in \mathbb{Z}$. Furthermore,

$$P(-m_{i+1}) = P(a_i v_i - m_i) \equiv P(-m_i) \equiv 0 \pmod{v_i},$$

the last congruence being the inductive step. Thus, $v_{i+1} = -P(-m_{i+1})/v_i \in \mathbb{Z}$, and $P(-m_{i+1}) = -v_i v_{i+1} \equiv 0 \pmod{v_{i+1}}$.

Bounding m_i and v_i: We now show that $v_i > 0$ for $i \geq 1$. Applying the conjugation in the field $\mathbb{Q}[\eta]$ to the expression for η_i from Lemma 6.3.1, we get

$$\bar{\eta}_i = \frac{-q_{i-2}\bar{\eta} + p_{i-2}}{q_{i-1}\bar{\eta} - p_{i-1}} = -\frac{q_{i-2}}{q_{i-1}} \cdot \frac{\bar{\eta} - \frac{p_{i-2}}{q_{i-2}}}{\bar{\eta} - \frac{p_{i-1}}{q_{i-1}}} < 0.$$

To justify the inequality, observe that the first factor is always negative, while the second is close to 1 for i large. Indeed, both its numerator and denominator approach $\bar{\eta} - \lim_{i\to\infty} p_i/q_i = \bar{\eta} - \eta \neq 0$. On the other hand, $\eta_i > 0$ as $i \geq 1$, and

$$0 < \eta_i - \bar{\eta}_i = \frac{m_i + \eta_O}{v_i} - \frac{m_i + \bar{\eta}_O}{v_i} = \frac{\eta_O - \bar{\eta}_O}{v_i}.$$

This is where our assumption $\eta_O > \bar{\eta}_O$ allows us to conclude that $v_i > 0$ for $i \geq 1$ (v_0 can be negative). The desired bound on v_i follows:

$$0 < v_i \leq v_i v_{i+1} = -P(-m_{i+1}) \leq -\min P(x).$$

This chain of inequalities gives $P(-m_i) < 0$, so that $-m_i$ lies between the two roots of $P(x)$:

$$-\eta_O < m_i < -\bar{\eta}_O. \qquad\qquad \blacksquare$$

For arithmetical application it is useful to relate the coefficient v_i to the field norm.

6.4.9 Proposition. *For η an algebraic integer and $i \geq 0$, we have $v_{i+1} = (-1)^{i+1} N(p_i - q_i \eta)$.*

Proof. As $\eta = \eta_O$ is an algebraic integer, we can find $f, g \in \mathbb{Z}$ for which $\eta^2 - f\eta + g = 0$, so that $\eta + \bar{\eta} = f$ and $\eta\bar{\eta} = g$. By Lemma 6.3.1,

$$\eta_{i+1} = \frac{-q_{i-1}\eta + p_{i-1}}{q_i\eta - p_i} = \frac{-q_{i-1}\eta + p_{i-1}}{q_i\eta - p_i} \cdot \frac{q_i\bar{\eta} - p_i}{q_i\bar{\eta} - p_i}$$

$$= \frac{(-p_ip_{i-1} - gq_iq_{i-1}) + p_{i-1}q_i\bar{\eta} + p_iq_{i-1}\eta}{\mathrm{N}(p_i - q_i\eta)}$$

$$= \frac{(-p_ip_{i-1} - gq_iq_{i-1} + fp_{i-1}q_i) + (-p_{i-1}q_i + p_iq_{i-1})\eta}{\mathrm{N}(p_i - q_i\eta)}$$

$$= \frac{(-p_ip_{i-1} - gq_iq_{i-1} + fp_{i-1}q_i) + (-1)^{i-1}\eta}{\mathrm{N}(p_i - q_i\eta)}$$

$$= \frac{(-1)^i(p_ip_{i-1} + gq_iq_{i-1} - fp_{i-1}q_i) + \eta}{(-1)^{i+1}\mathrm{N}(p_i - q_i\eta)}.$$

Comparing this with $\eta_{i+1} = (m_{i+1} + \eta)/v_{i+1}$ proves the proposition. We also get that $m_{i+1} = (-1)^i(p_ip_{i-1} + gq_iq_{i-1} - fp_{i-1}q_i)$. ∎

We conclude with a simple characterization of purely periodic quadratic numbers.

6.4.10 Theorem. *Let η be a quadratic number, which by Thm. 6.4.2 has a periodic continued fraction. That continued fraction is purely periodic if and only if $\eta > 1$ and $-1 < \bar{\eta} < 0$.*

Proof. Denote by l the period length of η. First, assume that η is purely periodic. By Prop. 6.4.4 (a), η and $\bar{\eta}$ are the roots of the polynomial $f(x) = q_{l-1}x^2 + (q_{l-2} - p_{l-1})x - p_{l-2}$. Since $\{a_i\}$ is a CF sequence, $1 \le a_l = a_0 = p_0 < \eta$. The recursion then shows that $p_i > 0$ for all $i \ge 0$, in addition to $q_i > 0$, which is always true.

The inequality $-1 < \bar{\eta} < 0$ holds if and only if $f(x)$ has a zero in the interval $(-1, 0)$. To show that it does, it's enough to show that $f(-1)$ and $f(0)$ have opposite signs. Indeed,

$$f(-1) = q_{l-1} - q_{l-2} + p_{l-1} - p_{l-2}$$

$$= a_{l-1}q_{l-2} + q_{l-3} - q_{l-2} + a_{l-1}p_{l-2} + p_{l-3} - p_{l-2}$$

$$= (a_{l-1} - 1)(q_{l-2} + p_{l-2}) + q_{l-3} + p_{l-3} > 0,$$

$$f(0) = -p_{l-2} < 0.$$

For the converse, assume that $\eta > 1$ and $-1 < \bar{\eta} < 0$. This is the case $i = 0$ of the bounds $\eta_i > 1, -1 < \bar{\eta}_i < 0$. The first inequality is true for all $i > 0$, since $\eta_i > \lfloor\eta_i\rfloor = a_i \ge 1$. To prove the second inequality, we argue inductively. If $\bar{\eta}_i < 0$, then $\bar{\eta}_i - a_i < -a_i \le -1$, hence $-1 < 1/(\bar{\eta}_i - a_i) = \bar{\eta}_{i+1} < 0$.

Rewrite the inequality $-1 < \bar{\eta}_i < 0$ as $0 < -\bar{\eta}_i = -1/\bar{\eta}_{i+1} - a_i < 1$, so that $a_i = \lfloor -1/\bar{\eta}_{i+1}\rfloor$. Since η is a quadratic number, we know that $\eta_k = \eta_{k+l}$ for all k big enough. From

$$a_{k-1} = \left\lfloor -\frac{1}{\bar{\eta}_k} \right\rfloor = \left\lfloor -\frac{1}{\bar{\eta}_{k+l}} \right\rfloor = a_{k+l-1}$$

we conclude that $\eta_{k-1} = a_{k-1} + 1/\eta_k = a_{k+l-1} + 1/\eta_{k+l} = \eta_{k+l-1}$. We keep lowering the index until we get $\eta_0 = \eta_l$: η is purely periodic. ∎

6.4.11 Example. Take $D \in \mathbb{N}$ not a perfect square. Observe that $\lfloor \sqrt{D} \rfloor + \sqrt{D} > \lfloor \sqrt{D} \rfloor \geq 1$, and $0 < \sqrt{D} - \lfloor \sqrt{D} \rfloor < 1$ by the definition of $\lfloor \sqrt{D} \rfloor$. Thus, $\lfloor \sqrt{D} \rfloor + \sqrt{D}$ satisfies the condition of Thm. 6.4.10 and its continued fraction is purely periodic:

$$\sqrt{D} + \lfloor \sqrt{D} \rfloor = [2\lfloor \sqrt{D} \rfloor, a_1, \ldots, a_{l-1}, 2\lfloor \sqrt{D} \rfloor, \ldots].$$

Subtracting the integer $\lfloor \sqrt{D} \rfloor$ only changes the first element in the continued fraction, giving

$$\sqrt{D} = [\lfloor \sqrt{D} \rfloor, \overline{a_1, \ldots 2\lfloor \sqrt{D} \rfloor}]. \qquad \square$$

Exercises

6.4.1. Prove Prop. 6.4.9 directly from the recursions (6.4.8).

6.4.2. Let η be a quadratic number.

(a) Give a necessary and sufficient condition on η for its period to start at a_1, i.e., $\eta = [a_0, \overline{a_1, \ldots, a_l}]$.

(b) Check that this condition is satisfied when η is a quadratic integer with $\eta > \bar{\eta}$.

(c) Give an example of an η that is not a quadratic integer, but with a period that starts with a_1.

6.4.3. Let $\sqrt{D} = [a_0, \overline{a_1, \ldots, a_l}]$. Prove that $a_i < 2a_0$ for $1 \leq i < l$. When computing the period of $\sqrt{D}$, we can stop as soon as we get $a_i = 2\lfloor \sqrt{D} \rfloor$.

6.4.4. (a) If $\eta = [\overline{a_0, a_1, \ldots, a_{l-2}, a_{l-1}}]$, then $-1/\bar{\eta} = [\overline{a_{l-1}, a_{l-2}, \ldots, a_1, a_0}]$.

(b) Let $\sqrt{D} = [a_0, \overline{a_1, \ldots, a_{l-1}, 2a_0}]$. Prove the mirror-symmetry of the sequence $a_1, \ldots, a_{l-1}$: $a_i = a_{l-i}$ for $1 \leq i \leq l-1$.

6.5 Computing Quadratic Continued Fractions

The following proposition is a simple variation on Prop. 6.4.6 that paves the way for effective computation of continued fractions of quadratic numbers.

6.5.1 Proposition. *Any quadratic number is of the form* $\eta = (m_0 + \sqrt{d})/v_0$, *for some* $d, m_0, v_0 \in \mathbb{Z}$ *satisfying* $v_0 \mid m_0^2 - d$. *Put* $v_{-1} = (m_0^2 - d)/v_0$. *The sequences* $\{m_i\}$ *and* $\{v_i\}$ *defined recursively by*

$$m_{i+1} = a_i v_i - m_i, \quad v_{i+1} = v_{i-1} + a_i(m_i - m_{i+1})$$

satisfy the following properties for $i \geq 1$:

(a) $\eta_i = (m_i + \sqrt{d})/v_i$;
(b) $m_i, v_i \in \mathbb{Z}$ *and* $v_i \mid m_i^2 - d$;
(c) $0 < v_i \leq d$ *and* $-\sqrt{d} < m_i < \sqrt{d}$.

Proof. The argument is parallel to the proof of Prop. 6.4.6, with $\eta_{\mathcal{O}}$ replaced by $\sqrt{d}$. It's worth pointing out that we're not requiring d to be square-free, and that v_0 can be negative. For example, $(1 - \sqrt{2})/3$ has to be rewritten as $(-3 + \sqrt{18})/(-9)$ to ensure that $v_{-1} \in \mathbb{Z}$. To find m_0, v_0 and d in general, follow Exer. 6.5.3.

The initial condition guarantees that the recursions produce m_i and v_i in $\mathbb{Z}$. What may be less obvious is that the recursion for v_i is the one in (6.4.8), with $P(x)$ replaced by $x^2 - d$. Indeed, starting with (6.4.8), we find that

$$v_{i+1} = -\frac{m_{i+1}^2 - d}{v_i} = -\frac{(a_i v_i - m_i)^2 - d}{v_i} = 2a_i m_i - a_i^2 v_i - \frac{m_i^2 - d}{v_i}$$
$$= a_i(m_i + m_i - a_i v_i) + v_{i-1} = v_{i-1} + a_i(m_i - m_{i+1}). \qquad \blacksquare$$

In principle, the formula $a_i = \lfloor (m_i + \sqrt{d})/v_i \rfloor$ combined with the recursions of the proposition gives an algorithm for determining the continued fraction of η. Both computationally and philosophically, it would be better not to have to deal with $\sqrt{d}$. Indeed, requiring the foreknowledge of the decimal expansion of $\sqrt{d}$ would defeat the purpose of the algorithm, which is to efficiently describe irrational numbers purely in terms of integers. Fortunately, it suffices to replace $\sqrt{d}$ by its crudest approximation, $\lfloor \sqrt{d} \rfloor$.

6.5.2 Proposition. *For all* $i \geq 1$ *we have* $a_i = \lfloor (m_i + \lfloor \sqrt{d} \rfloor)/v_i \rfloor$. *In addition,* $a_0 = \lfloor (m_0 + \lfloor \sqrt{d} \rfloor)/v_0 \rfloor$ *or* $\lfloor (m_0 + \lfloor \sqrt{d} \rfloor)/v_0 \rfloor - 1$. *The second possibility occurs if and only if* $v_0 < 0$ *and* $v_0 \mid m + \lfloor \sqrt{d} \rfloor$.

Proof. We leave the case $i = 0$ as an exercise, and assume that $i \geq 1$. By Prop. 6.5.1 (c), $a_i v_i - m_i = m_{i+1} \leq \lfloor \sqrt{d} \rfloor$. Dividing by $v_i > 0$ gives the first inequality in the following chain:

$$a_i \leq \frac{m_i + \lfloor \sqrt{d} \rfloor}{v_i} < \frac{m_i + \sqrt{d}}{v_i} < a_i + 1.$$

This just means that $a_i = \lfloor (m_i + \lfloor \sqrt{d} \rfloor)/v_i \rfloor$. $\qquad \blacksquare$

For convenience, we extract from the proof of Prop. 6.5.2 an algorithm for finding the continued fraction of a quadratic number η:

THE CONTINUED FRACTION ALGORITHM FOR QUADRATIC NUMBERS

1. INPUT: Quadratic number $\eta \in \mathbb{R}$. Put $\eta_0 := \eta = (m_0 + \sqrt{d})/v_0$ with $v_0 \mid d - m_0^2$, as in Exer. 6.5.3. Put $v_{-1} := (d - m_0^2)/v_0$, $s := \lfloor \sqrt{d} \rfloor$, and set the counter $i := 0$.

2. If $(m_i, v_i) = (m_r, v_r)$ for some $0 \leq r < i$, OUTPUT:

$$\eta = [a_0, \ldots, a_{r-1}, \overline{a_r, \ldots, a_{i-1}}].$$

3. Otherwise compute, in the following order:

$$a := \lfloor (m_i + s)/v_i \rfloor$$

$$a_i := \begin{cases} a - 1 & \text{if } i = 0, v_0 < 0 \text{ and } v_0 \mid m_0 + s \\ a & \text{otherwise} \end{cases}$$

$$m_{i+1} := a_i v_i - m_i$$

$$v_{i+1} := v_{i-1} + a_i(m_i - m_{i+1}).$$

Increase the counter i by 1 and go to Step 2.

6.5.3 Example. To find the continued fraction for $\sqrt{223}$, we put $d = 223, m_0 = 0, v_0 = 1, s = 14$, and fill in the following table using the algorithm, keeping an eye out for periodicity:

i	-1	0	1	2	3	4	5
a_i	–	14	1	13	1	28	1
m_i	–	0	14	13	13	14	14
v_i	223	1	27	2	27	1	27

As $m_5 = m_1$ and $v_5 = v_1$, we get $\sqrt{223} = [14, \overline{1, 13, 1, 28}]$.

An interesting piece of information lurks in this table. Denote by p_i/q_i the ith convergent of $\sqrt{223}$. Then $\{p_i\}$ and $\{q_i\}$ are both strictly increasing sequences of integers. By Prop. 6.4.9, however, the only possible values for $N(p_i - q_i\sqrt{223}) = (-1)^{i+1}v_{i+1}$ are 1, -27, and 2. $\qquad \square$

Exercises

6.5.1. Apply the Continued Fraction Algorithm for Quadratic Numbers to:
(a) $\sqrt{163}$; (b) $(4 + \sqrt{89})/5$; (c) $(2 - \sqrt{7})/3$; (d) $\sqrt{319}$.

6.5.2. Prove the claim about a_0 made in Prop. 6.5.2. Give an example of an η for which $a_0 = \lfloor (m_0 + \sqrt{d})/v_0 \rfloor - 1$.

6.5.3. Let $\eta = (x + y\sqrt{z})/w$ with $x, y, z, w \in \mathbb{Z}, w \neq 0$, and $z > 0$ not a perfect square. Put $g = w/\gcd(w, y^2z - x^2)$, $d = g^2y^2z$, $m_0 = \mathrm{sgn}(y)gx$, and $v_0 = \mathrm{sgn}(y)gw$. Show that $\eta = (m_0 + \sqrt{d})/v_0$ with $v_0 \mid d - m_0^2$. Prove that d is the smallest possible such value.

6.6 Approximation by Convergents

We hinted at the beginning of this chapter that a real number is efficiently approximated by the convergents of its continued fraction. In this section we make that claim precise. We will repeatedly use the simple observation that, for $a, p \in \mathbb{Z}$ and $b, q \in \mathbb{N}$,

$$\frac{a}{b} \neq \frac{p}{q} \text{ implies } \left| \frac{a}{b} - \frac{p}{q} \right| = \frac{|aq - bp|}{bq} \geq \frac{1}{bq},$$

as $aq - bp$ is a non-zero integer. We assume throughout that $\eta = [a_0, a_1, \dots]$ is irrational, with convergents p_i/q_i.

6.6.1 Proposition. *Let $a, b \in \mathbb{Z}$ with $b > 0$ and $\gcd(a, b) = 1$. If $\left| \eta - \frac{a}{b} \right| < \left| \eta - \frac{p_i}{q_i} \right|$, then $b > q_i$.*

Put plainly, if we want to improve the approximation of η given by a convergent, we need a fraction with bigger denominator.

Proof. For simplicity, let's assume that i is odd, so that

$$p_{i+1}/q_{i+1} < \eta < p_i/q_i.$$

The even case is similar, with the inequalities reversed. By assumption, a/b is in the interval centered at η with one endpoint at p_i/q_i. We leave the right half of that interval as an exercise, and consider only the case $a/b \in (2\eta - p_i/q_i, \eta)$. Since the p_i/q_i get steadily closer to η (Exer. 6.3.2), p_{i+1}/q_{i+1} is in the same interval (see Fig. 6.1).

The length of $(2\eta - p_i/q_i, \eta)$ is therefore greater than the distance between p_{i+1}/q_{i+1} and a/b. This gives the second inequality in

$$(6.6.2) \qquad \frac{1}{bq_{i+1}} \leq \left| \frac{a}{b} - \frac{p_{i+1}}{q_{i+1}} \right| < \left| \eta - \frac{p_i}{q_i} \right| < \frac{1}{q_iq_{i+1}}.$$

We conclude that $b > q_i$. ∎

How can we tell whether a fraction is a convergent of η? The following proposition gives a sufficient condition.

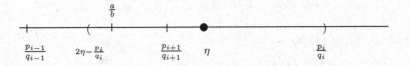

Fig. 6.1 Illustration for Props. 6.6.1 and 6.6.3 (odd i).

6.6.3 Proposition. *For $a, b \in \mathbb{Z}$ with $b > 0$ and $\gcd(a, b) = 1$,*

$$\left| \eta - \frac{a}{b} \right| < \frac{1}{2b^2} \text{ implies } a = p_i, b = q_i \text{ for some } i \geq 0.$$

Proof. We argue by contradiction. If $a/b \neq p_k/q_k$ for all k, there exists a unique i such that a/b is between p_{i-1}/q_{i-1} and p_{i+1}/q_{i+1}. We assume that i is odd; the even case is analogous. The situation is pictured in Fig. 6.1, with the caveat that, unlike in the preceding proof, a/b can be on either side of $2\eta - p_i/q_i$.

We have the following chain of inequalities, the last of which is apparent from Fig. 6.1:

$$\left| \eta - \frac{p_i}{q_i} \right| < \frac{1}{q_i q_{i+1}} = \frac{b}{q_i} \cdot \frac{1}{b q_{i+1}} \leq \frac{b}{q_i} \left| \frac{p_{i+1}}{q_{i+1}} - \frac{a}{b} \right| < \frac{b}{q_i} \left| \eta - \frac{a}{b} \right|.$$

Combining this with the triangle inequality, we get

$$\frac{1}{b q_i} \leq \left| \frac{a}{b} - \frac{p_i}{q_i} \right| \leq \left| \eta - \frac{a}{b} \right| + \left| \eta - \frac{p_i}{q_i} \right| < \left| \eta - \frac{a}{b} \right| \left(1 + \frac{b}{q_i} \right).$$

Only now do we apply the assumption $|\eta - a/b| < 1/2b^2$:

$$\frac{1}{b q_i} < \left| \eta - \frac{a}{b} \right| \left(1 + \frac{b}{q_i} \right) < \frac{1}{2b^2} \cdot \frac{q_i + b}{q_i}.$$

Thus $b < q_i$.

On the other hand, a/b lies in the interval $(p_{i-1}/q_{i-1}, \eta)$. Arguing similarly to (6.6.2) gives $q_i < b$. That contradiction shows that a/b is a convergent. ∎

The following corollary is the key to our application of continued fractions to real quadratic fields.

6.6.4 Corollary. *Let $P(x) = x^2 - fx + g$ be a polynomial with coefficients in $\mathbb{Z}$, of discriminant $D = f^2 - 4g > 0$. Assume that its roots satisfy $\eta > 0 > \bar{\eta}$, and let p_i/q_i be the convergents of the continued fraction for η. For relatively prime $a, b \in \mathbb{N}$, the following implication holds:*

If $\left|a^2 - fab + gb^2\right| < \dfrac{\sqrt{D}}{2}$, *then* $a = p_i, b = q_i$ *for some* $i > 0$.

Proof. Dividing the bound by b^2 gives

$$\left|P\left(\frac{a}{b}\right)\right| = \left|\left(\frac{a}{b}\right)^2 - f\left(\frac{a}{b}\right) + g\right| < \frac{\sqrt{D}}{2b^2}.$$

We distinguish two cases.

Case 1: $\eta < a/b$. Since $P'(\eta) = \sqrt{D} > 0$, and since the parabola $y = P(x)$ is above its tangent line at η, we have that $0 < P'(\eta)(x - \eta) < P(x)$ when $\eta < x$. Putting $x = a/b$ and dividing by $P'(\eta)$, we get

$$\left|\frac{a}{b} - \eta\right| = \frac{a}{b} - \eta < \frac{P(a/b)}{\sqrt{D}} = \frac{|P(a/b)|}{\sqrt{D}} < \frac{\sqrt{D}/2b^2}{\sqrt{D}} = \frac{1}{2b^2},$$

so that $a = p_i$, $b = q_i$ for some i, by Prop. 6.6.3.

Case 2: $0 < a/b < \eta$. By our assumption $\eta > 0 > \bar{\eta}$, the polynomial $Q(x) = -x^2 P(1/x) = -gx^2 + fx - 1$ has roots $1/\eta > 0 > 1/\bar{\eta}$ and satisfies $Q'(1/\eta) = \sqrt{D} > 0$. Moreover, its leading coefficient $-g = -\eta\bar{\eta}$ is positive, so $Q(x)$ shares the salient properties of $P(x)$. If $0 < x < \eta$, then $1/\eta < 1/x$, and the same reasoning as in Case 1 gives

$$\left|\frac{1}{x} - \frac{1}{\eta}\right| < \frac{Q(1/x)}{\sqrt{D}} = \frac{-P(x)}{x^2\sqrt{D}} = \frac{|P(x)|}{x^2\sqrt{D}},$$

the last equality because $P(x) < 0$ for $0 < x < \eta$. For $x = a/b$ this becomes

$$\left|\frac{b}{a} - \frac{1}{\eta}\right| < \frac{|P(a/b)|}{(a^2/b^2)\sqrt{D}} < \frac{\sqrt{D}/2b^2}{(a^2/b^2)\sqrt{D}} = \frac{1}{2a^2}.$$

By Prop. 6.6.3 again, b/a is a convergent of $1/\eta$. We see from Exer. 6.2.4 that $b/a = q_i/p_i$ for some i, as desired. ∎

6.6.5 Example. Let $F = \mathbb{Q}[\sqrt{223}]$. In Ex. 5.4.12 we showed that $\mathrm{Cl}(F) \cong \mathbb{Z}/3\mathbb{Z}$, provided that $x^2 - 223y^2 = \pm 3$ has no integer solutions. That is, indeed, the case. If not, assume $a, b \in \mathbb{N}$ satisfy the equation, and apply Cor. 6.6.4 with $\eta = \sqrt{223}$: $\left|a^2 - 223b^2\right| = 3 < \sqrt{4 \cdot 223}/2$ gives that $a = p_i, b = q_i$, for some convergent p_i/q_i of $\sqrt{223}$. In Ex. 6.5.3 we found that $p_i^2 - 223q_i^2$ can only be 1, -27, and 2. Thus, $x^2 - 223y^2 = \pm 3$ has no integer solutions. □

Exercises

6.6.1. Finish the proof of Prop. 6.6.1 by handling the case $\eta < a/b < p_i/q_i$.

6.6.2. Verify the claims made about the polynomial $Q(x)$ in the proof of Cor. 6.6.4. Where do we use the assumption $\eta > 0 > \bar{\eta}$?

6.6.3. Find relatively prime solutions $x, y \in \mathbb{Z}$ to each of the following equations, or prove that there isn't one:

(a) $x^2 - 58y^2 = 7$, $x^2 - 58y^2 = 5$
(b) $x^2 - 89y^2 = -5$, $x^2 - 89y^2 = 40$
(c) $x^2 - 43y^2 = 36$
(d) $x^2 - 73y^2 = 72$
(e) $x^2 - xy - 48y^2 = -14$

6.6.4.* Compute $\mathrm{Cl}(\mathbb{Q}[\sqrt{D}])$ for $D \in \{82, 401, 469, 577\}$. Use Cor. 6.6.4 to decide when ideals are principal, as in Ex. 6.6.5.

6.6.5. Conclude Exer. 5.4.6 (c): show that $\mathrm{Cl}(\mathbb{Q}[\sqrt{399}]) \cong \mathbb{Z}/4\mathbb{Z} \times \mathbb{Z}/2\mathbb{Z}$.

6.7 The Group of Units of a Real Quadratic Field

So far, we've looked at continued fractions as a tool for approximating real numbers. But they have another, unexpected use: determining the group of units in the ring of integers $\mathcal{O} = \mathbb{Z}[\delta]$ of a real quadratic field F. In Exer. 4.2.4 we saw that there is some leeway in choosing δ. In the context of real quadratic fields, however, there is one preferred choice.

6.7.1 Proposition. *There exists precisely one Δ such that $\mathcal{O} = \mathbb{Z}[\Delta]$ and the continued fraction of Δ is purely periodic: $\Delta > 1$ and $-1 < \bar{\Delta} < 0$.*

Proof. After possibly replacing δ by $\bar{\delta}$, we may assume that $\delta > \bar{\delta}$. Any δ' with $\mathcal{O} = \mathbb{Z}[\delta']$ must be of the form $\delta' = a \pm \delta$ for some $a \in \mathbb{Z}$. Assume that δ' is purely periodic, which by Thm. 6.4.10 means that

$$a \pm \delta > 1 \text{ and } -1 < a \pm \bar{\delta} < 0.$$

In particular, $\pm \delta > -a > \pm \bar{\delta}$, so our condition $\delta > \bar{\delta}$ forces the sign to be plus, i.e., $\delta' = a + \delta$. The second inequality then gives $a + 1 > -\bar{\delta} > a$, in other words, $a = \lfloor -\bar{\delta} \rfloor$.

If there is a periodic δ', it thus must be $\Delta = \lfloor -\bar{\delta} \rfloor + \delta$. It remains to check that $\Delta > 1$. The smallest possible discriminant of a real quadratic field is $D_{\mathbb{Q}[\sqrt{5}]} = 5$. Then $2 < \sqrt{D_F} = \delta - \bar{\delta} < \delta + \lfloor -\bar{\delta} \rfloor + 1$, whence $1 < \lfloor -\bar{\delta} \rfloor + \delta = \Delta$. ∎

For the remainder of this section, we fix the usual notations in reference to Δ. Thus, t and n are defined by $\Delta^2 - t\Delta + n = 0$, while a_i, p_i, q_i and Δ_i are the quantities associated with the continued fraction of Δ. Since Δ is an algebraic integer, $\Delta_{\mathcal{O}} = \Delta$. As in the proof of Prop. 6.4.6, we put

$$\Delta_i = \frac{m_i + \Delta}{v_i} \text{ with } v_i, m_i \in \mathbb{Z} \text{ and } v_i \mid m_i^2 + tm_i + n.$$

Finally, l stands for the length of the period of Δ, which we call the **period length of F**.

6.7.2 Proposition. *For $i > 0$, $v_i = 1$ if and only if i is a multiple of l.*

Proof. Assume $v_i = 1$, so that $\Delta_i = m_i + \Delta$. As both Δ and Δ_i are purely periodic, Thm. 6.4.10 implies that we have $-1 < \bar{\Delta} < 0$ and $-1 < m_i + \bar{\Delta} < 0$. Since $m_i \in \mathbb{Z}$, this is only possible if $m_i = 0$. Then $\Delta_i = \Delta$ and i is a multiple of the period length.

Conversely, if $i = lk$ for some k, we have $\Delta = \Delta_{lk} = (m_{lk} + \Delta)/v_{lk}$. Comparing coefficients gives $v_{lk} = 1$. ∎

The unit group in a real quadratic field turns out to essentially consist of powers of a single unit.

6.7.3 Definition. *A **fundamental unit** of $\mathcal{O}$ (or F) is a unit $\varepsilon_F \in \mathcal{O}^\times$ satisfying the following conditions:*

(a) For any $\varepsilon \in \mathcal{O}^\times$, $\varepsilon = \pm\varepsilon_F^k$ for some $k \in \mathbb{Z}$, and
(b) $\varepsilon_F > 1$.

Condition (a) means that $\mathcal{O}^\times = \pm\langle\varepsilon_F\rangle$. Here $\langle\varepsilon_F\rangle$ is the multiplicative infinite cyclic group generated by ε_F (don't confuse it with the principal ideal generated by ε_F, which is just $\mathcal{O}$). Condition (b) guarantees that ε_F is unique, if it exists. The following theorem shows how to construct it.

6.7.4 Theorem. *Let $\mathcal{O} = \mathbb{Z}[\delta]$ with $\delta > \bar{\delta}$. Let p_i'/q_i' be the convergents of δ, and l the period length of F.*

Put $\varepsilon_1 = p_{l-1}' - q_{l-1}'\delta$. Any unit $\varepsilon \in \mathcal{O}^\times$ is of the form $\varepsilon = \pm\varepsilon_1^k$ for some $k \in \mathbb{Z}$. The fundamental unit ε_F is then $\pm\varepsilon_1$ or $\pm\bar{\varepsilon}_1$, whichever is greater than 1.

Proof. By Prop. 6.4.9, ε_1 is a unit. As $\delta > \bar{\delta}$, there is an $a \in \mathbb{Z}$ such that $\delta = \Delta + a$. By Exer. 6.2.3, the convergents of δ and Δ are related by $p_i' = aq_i + p_i$ and $q_i' = q_i$. We compute

$$\varepsilon_1 = p_{l-1}' - q_{l-1}'\delta = p_{l-1} + aq_{l-1} - q_{l-1}\delta = p_{l-1} - q_{l-1}(\delta - a)$$
$$= p_{l-1} - q_{l-1}\Delta.$$

We henceforth assume, without loss of generality, that $\delta = \Delta$.

A unit ε is a power of ε_1 up to sign if the same is true for $-\varepsilon$, or for $\bar{\varepsilon} = \pm\varepsilon^{-1}$. This allows us to restrict to the case $\varepsilon = a - b\Delta$ for $a, b \in \mathbb{Z}_{>0}$. Indeed, if $a < 0$, replace ε by $-\varepsilon$. Then, if $b < 0$, replace ε by $\bar{\varepsilon} = a - b\bar{\Delta} = (a - bt) - (-b)\Delta$. Since Δ is purely periodic, $t > 0$ and $a - bt$ is still positive.

Since $D_F \geq D_{\mathbb{Q}[\sqrt{5}]} = 5$, we have

$$\left|a^2 - tab + nb^2\right| = |N(a - b\Delta)| = 1 < \sqrt{5}/2 \leq \sqrt{D_F}/2.$$

We conclude from Cor. 6.6.4 that $a = p_i, b = q_i$ for some $i > 0$. Prop. 6.4.9 then gives

$$v_{i+1} = (-1)^{i+1}N(p_i - q_i\Delta) = (-1)^{i+1}N\varepsilon = \pm1.$$

As $v_{i+1} > 0$ by the proof of Prop. 6.4.6, we conclude that $v_{i+1} = 1$, and therefore, by Prop. 6.7.2, that $i + 1 = kl$, a multiple of the period.

We have shown that every $a - b\Delta \in \mathcal{O}^\times$ with $a, b \in \mathbb{Z}_{\geq 0}$ must be of the form

$$\varepsilon_k = p_{kl-1} - q_{kl-1}\Delta, \text{ for some } k \geq 0.$$

To prove the theorem it suffices to verify that $\varepsilon_{k+1}/\varepsilon_k = \varepsilon_1$. By Lemma 6.3.1,

$$-\Delta_{i+1} = \frac{p_{i-1} - q_{i-1}\Delta}{p_i - q_i\Delta}.$$

We form a telescoping product:

$$(-1)^l \Delta_{i+1}\Delta_{i+2}\cdots\Delta_{i+l} = \frac{p_{i-1} - q_{i-1}\Delta}{p_i - q_i\Delta} \cdot \frac{p_i - q_i\Delta}{p_{i+1} - q_{i+1}\Delta} \cdots \frac{p_{i+l-2} - q_{i+l-2}\Delta}{p_{i+l-1} - q_{i+l-1}\Delta}$$

$$= \frac{p_{i-1} - q_{i-1}\Delta}{p_{i+l-1} - q_{i+l-1}\Delta}.$$

When $i = kl$, we get

$$(-1)^l \Delta_{kl+1}\Delta_{kl+2}\cdots\Delta_{(k+1)l} = \frac{p_{kl-1} - q_{kl-1}\Delta}{p_{(k+1)l-1} - q_{(k+1)l-1}\Delta} = \frac{\varepsilon_k}{\varepsilon_{k+1}}.$$

When $k = 0$, this in turn becomes

$$(-1)^l \Delta_1 \Delta_2 \cdots \Delta_l = \frac{p_{-1} - q_{-1}\Delta}{p_{l-1} - q_{l-1}\Delta} = \frac{1}{p_{l-1} - q_{l-1}\Delta} = \frac{1}{\varepsilon_1}.$$

Periodicity implies $\Delta_{kl+1}\Delta_{kl+2}\cdots\Delta_{(k+1)l} = \Delta_1\Delta_2\ldots\Delta_l$. Consequently, $\varepsilon_k/\varepsilon_{k+1} = 1/\varepsilon_1$, as desired. ∎

6.7.5 Example. We illustrate Thm. 6.7.4 by finding the fundamental unit of $\mathbb{Z}[\sqrt{223}]$. We compute the convergents of the continued fraction we found in Ex. 6.5.3, $\sqrt{223} = [14, \overline{1, 13, 1, 28}]$. It suffices to go up to the end of the period:

i	-2	-1	0	1	2	3	4	$\cdots$
a_i	$-$	$-$	14	1	13	1	28	$\cdots$
p_i	0	1	14	15	209	224	6481	$\cdots$
q_i	1	0	1	1	14	15	434	$\cdots$

Theorem 6.7.4 shows that, up to sign, $\mathbb{Z}[223]^\times$ is generated by $\varepsilon_1 = p_3 - q_3\sqrt{223} = 224 - 15\sqrt{223} = 0.0022\ldots$. The fundamental unit must be greater than 1, so we obtain it as $\bar{\varepsilon}_1$:

$$\varepsilon_{\mathbb{Q}[\sqrt{223}]} = p_3 + q_3\sqrt{223} = 224 + 15\sqrt{223}. \square$$

In general, we read off the fundamental unit from the column under the next-to-last element of the period of the continued fraction of δ.

Exercises

6.7.1. Show that the unique purely periodic generator of $\mathcal{O}$ is

$$\Delta = \left\lfloor \frac{\sqrt{D_F} - D_F}{2} \right\rfloor + \frac{D_F + \sqrt{D_F}}{2}.$$

6.7.2. Show that i is a multiple of the period length of F if and only if $\delta_i \in \mathcal{O}$, where $\mathcal{O} = \mathbb{Z}[\delta]$.

6.7.3. Prove that there is at most one unit satisfying the conditions (a) and (b) of Def. 6.7.3. Check the last claim of Thm. 6.7.4 by showing that precisely one of $\pm\varepsilon_1, \pm\bar{\varepsilon}_1$ is greater than 1.

6.7.4. Find the fundamental unit and the period of $\mathbb{Q}[\sqrt{D}]$, for $D \in \{7, 13, 17, 41, 70, 58, 102\}$.

6.7.5.* We know a priori that $N\varepsilon_F = \pm1$. How can we determine the sign?

(a) Prove that $N\varepsilon_F = 1$ if and only if no unit in $\mathcal{O}$ has norm -1.
(b) Show that $N\varepsilon_F = -1$ if and only if the period length of F is odd.
(c) If D_F is divisible by a positive prime $p \equiv 3 \pmod 4$, then $N\varepsilon_F = 1$.
(d) If $D_F = p$, a positive prime, then $N\varepsilon_F = -1$.

Orders

6.7.6. Let F be a quadratic field, real or imaginary, and let $\mathcal{O}_c \neq \mathcal{O}$ be the order of conductor $c > 1$. Let $\varphi : F_1^\times(c) \to (\mathcal{O}/c\mathcal{O})^\times/(\mathbb{Z}/c\mathbb{Z})^\times$ be the homomorphism defined in Exer. 5.1.12 (c).

(a) If F is imaginary, show that $\mathcal{O}_c^\times = \{\pm1\}$.
(b) Assume from now on that F is real. Let $\varepsilon_1 = p_{l-1} - q_{l-1}\Delta = \pm\varepsilon_F^{\pm1}$ be the unit constructed in the proof of Thm. 6.7.4. Show that $\varphi(\mathcal{O}^\times) = \langle\varphi(\varepsilon_1)\rangle$.
(c) Let t be the order of $\varphi(\varepsilon_1)$ in $(\mathcal{O}/c\mathcal{O})^\times/(\mathbb{Z}/c\mathbb{Z})^\times$. Show that t is the smallest positive integer for which $c \mid q_{tl-1}$. Without this group-theoretic argument, it is not a priori obvious that such a t exists.
(d) Show that $\mathcal{O}_c^\times = \pm\langle\varepsilon_F^s\rangle$, and that $s = [\mathcal{O}^\times : \mathcal{O}_c^\times]$.

Chapter 7
Quadratic Forms

7.1 Motivation

In this final chapter we go back to the late-eighteenth-century roots of algebraic number theory. Its fathers, Lagrange, Legendre, and Gauss, had none of the algebraic machinery we have used. They discovered the ideal class group solely by studying certain $\mathbb{Z}$-valued functions on the standard lattice $\Lambda_0 = \{[\begin{smallmatrix} x \\ y \end{smallmatrix}] : x, y \in \mathbb{Z}\}$.

7.1.1 Definition. *A* **form** *is a function* $q : \Lambda_0 \to \mathbb{Z}$ *given by* $q(x, y) = ax^2 + bxy + cy^2$, *for some* $a, b, c \in \mathbb{Z}$.

Even though Λ_0 consists of *column* vectors, we write $q(x, y)$ instead of the correct but cumbersome $q([\begin{smallmatrix} x \\ y \end{smallmatrix}])$. We reserve the more common term "quadratic form" for forms that are nontrivial, in the sense described in the next section.

7.1.2 Example. We can rephrase Fermat's question from the beginning of the book as follows: which primes are possible values of the form $x^2 + y^2$? In Ex. 5.4.12 we needed to decide whether the prime ideal P_3 of $\mathbb{Z}[\sqrt{223}]$ is principal. That boiled down to determining whether the form $x^2 - 223y^2$ could take the value 3 or -3.

Note that $x^2 + y^2$ computes the norm of $x + yi \in \mathbb{Z}[i]$ in terms of its coordinates relative to the basis $\{1, i\}$ of $\mathbb{Z}[i]$. A similar interpretation holds for $x^2 - 223y^2$ and the basis $\{1, \sqrt{223}\}$ of $\mathbb{Z}[\sqrt{223}]$. □

7.1.3 Example. The preceding example generalizes to any ideal with a chosen basis. Let $\mathscr{I} = \mathbb{Z}\alpha + \mathbb{Z}\beta$ be a fractional ideal in a quadratic field F, with a choice of an *ordered* basis (α, β). The expression

$$q_{\mathscr{I}, \alpha, \beta}(x, y) = \frac{\mathrm{N}(x\alpha - y\beta)}{\mathrm{N}\mathscr{I}} = \frac{\mathrm{N}\alpha}{\mathrm{N}\mathscr{I}}x^2 - \frac{\mathrm{Tr}(\bar{\alpha}\beta)}{\mathrm{N}\mathscr{I}}xy + \frac{\mathrm{N}\beta}{\mathrm{N}\mathscr{I}}y^2$$

M. Trifković, *Algebraic Theory of Quadratic Numbers*, Universitext,   131
DOI 10.1007/978-1-4614-7717-4_7, © Springer Science+Business Media New York 2013

has coefficients in $\mathbb{Z}$ by Exer. 4.6.5. It thus defines a form, called the form **associated** to $(\mathscr{I}, \alpha, \beta)$. It is is invariant under scaling by half the elements of $F^\times$:

$$(7.1.4) \qquad q_{\gamma\mathscr{I},\gamma\alpha,\gamma\beta}(x,y) = q_{\mathscr{I},\alpha,\beta}(x,y), \text{ when } \mathrm{N}\gamma > 0.$$

This suggests a connection between forms and ideal classes that is the theme of this chapter. The link is subtler than it appears: $q_{\mathscr{I},\alpha,\beta}$ depends not only on (the ideal class of) $\mathscr{I}$, but also on the choice of ordered basis (α, β). Any other ordered basis of $\mathscr{I}$ is of the form $(\alpha', \beta') = (m\alpha + l\beta, k\alpha + j\beta)$ for $\left[\begin{smallmatrix} j & k \\ l & m \end{smallmatrix}\right] \in \mathrm{GL}_2(\mathbb{Z})$. The forms associated with the two bases are related by

$$(7.1.5) \qquad \begin{aligned} q_{\mathscr{I},\alpha',\beta'}(x,y) &= \frac{\mathrm{N}(x(m\alpha + l\beta) - y(k\alpha + j\beta))}{\mathrm{N}\mathscr{I}} \\ &= \frac{\mathrm{N}((mx - ky)\alpha - (-lx + jy)\beta)}{\mathrm{N}\mathscr{I}} \\ &= q_{\mathscr{I},\alpha,\beta}(mx - ky, -lx + jy). \end{aligned}$$

Together with (7.1.4), this shows that the set of values of $q_{\mathscr{I},\alpha,\beta}$ depends only on the ideal class of $\mathscr{I}$, not on basis chosen (at least for imaginary quadratic F, where the norm is always positive).

As a convenient normalization, we say that the form $q_{\mathscr{I},\alpha',\beta'}$ is obtained from $q_{\mathscr{I},\alpha,\beta}$ by a **linear change of variables** $\left[\begin{smallmatrix} j & k \\ l & m \end{smallmatrix}\right] \in \mathrm{GL}_2(\mathbb{Z})$. If we don't want to specify the matrix, we say that the two forms are **equivalent**. They are **properly equivalent** if, in addition, we can take $\det\left[\begin{smallmatrix} j & k \\ l & m \end{smallmatrix}\right] = 1$. In Sec. 7.5 we will study proper equivalence in detail. $\qquad\square$

We can simplify any form $q(x,y) = ax^2 + bxy + cy^2$ by repeated linear changes of variables of two types:

$$(T^k q)(x,y) = q(x - ky, y) = ax^2 + (b - 2ak)xy + (ak^2 - bk + c)y^2,$$

one for each $k \in \mathbb{Z}$, and

$$(Sq)(x,y) = q(y, -x) = cx^2 - bxy + ay^2.$$

7.1.6 Proposition. *For any form there is a sequence of linear changes of variables, each of type T^k or S, which produces a form $ax^2 + bxy + cy^2$ satisfying*

$$(7.1.7) \qquad |b| \le |a| \le |c|, \text{ and } b \ge 0 \text{ when } |a| = |b| \text{ or } |a| = |c|.$$

Proof. We apply the following algorithm, due to Legendre:

DEFINITE REDUCTION ALGORITHM

1. INPUT: coefficients $a, b, c \in \mathbb{Z}$. Put $q(x,y) := ax^2 + bxy + cy^2$.

2. The variant of the division algorithm, Exer. 1.1.2, gives $k, r \in \mathbb{Z}$ with $b = k(2a) + r$ and $|r| \leq |a|$. The change of variables $q := T^k q$ for this k results in a form with $|b| \leq |a|$.

3. If $|a| \leq |c|$, go to Step 5.

4. Put $q := Sq$ to produce a form with $|a| < |c|$, and go back to Step 2.

5. If $b = -|a|$, put $q := T^{\operatorname{sgn} a} q$ to replace $ax^2 - |a| xy + cy^2$ with $ax^2 + (-|a| + 2 |a|) xy + cy^2 = ax^2 + |a| xy + cy^2$.

6. If $b < 0$ and $|a| = |c|$, put $q := Sq$ to replace $ax^2 + bxy + cy^2$ with $cx^2 - bxy + ay^2$.

7. OUTPUT: $q(x, y)$.

The positive integer $|b|$ either decreases (in Step 2), or stays constant (in Step 4). As the algorithm never applies Step 4 twice in a row, $|b|$ keeps decreasing and we eventually exit the loop made of Steps 2–4. Finally, if $b < 0$ in the two special cases in (7.1.7), Step 4 or Step 5 changes b to $-b$. ∎

7.1.8 Example. Let $F = \mathbb{Q}[\sqrt{-47}]$, so that $\mathcal{O} = \mathbb{Z}[\delta]$ for $\delta^2 - \delta + 12 = 0$. The form corresponding to the ideal $I = \mathbb{Z} \cdot 27 + \mathbb{Z}(47 + \delta)$ with the indicated ordered basis is $q_{I,27,47+\delta}(x, y) = 27x^2 - 95xy + 84y^2$. We apply the Definite Reduction Algorithm to this form:

$$27x^2 - 95xy + 84y^2 \xrightarrow{T^{-2}} 27x^2 + 13xy + 2y^2$$

$$\xrightarrow{S} 2x^2 - 13xy + 27y^2 \xrightarrow{T^{-3}} 2x^2 - xy + 6y^2.$$

Check that $2x^2 - xy + 6y^2 = q_{P_2,2,\delta}(x, y)$, and that I and $P_2 = \mathbb{Z} \cdot 2 + \mathbb{Z}\delta$ lie in the same ideal class. □

7.1.9 Example. Let's list all forms $ax^2 + bxy + cy^2$ which have **discriminant** $b^2 - 4ac = -47$, have $a > 0$ (and therefore $c > 0$), and satisfy condition (7.1.7), which now reads $|b| \leq a \leq c$, and $b \geq 0$ when $a = |b|$ or $a = c$. Such a form will be called **reduced**. From

$$47 = 4ac - b^2 \geq 4 |b|^2 - b^2 = 3b^2$$

we deduce that $|b| < \sqrt{47/3}$ and hence $|b| \leq 3$. For each b, we find the finitely many possibilities for a and c by factoring $b^2 + 47 = 4ac$. This equality forces b to be odd.

There are no reduced forms with $b = \pm 3$, because such a form would satisfy $ac = 14$ and $3 \leq a \leq c$. If $b = \pm 1$, then $ac = 12$ and we have the following possibilities:

(a) $a = 1, c = 12$: we get a single form $x^2 + xy + 12y^2$. We discard $x^2 - xy + 12y^2$ since $|b| = a$ in this case.

(b) $a = 2, c = 6$: we get two forms $2x^2 \pm xy + 6y^2$.
(c) $a = 3, c = 4$: we again get two forms $3x^2 \pm xy + 4y^2$.

Check that no two of these forms are properly equivalent.   □

7.1.10 Example. There are five reduced forms of discriminant -47, and there are also five ideal classes in $\mathrm{Cl}(\mathbb{Q}[\sqrt{-47}]) = \langle [P_2] \rangle$, as we calculated in Ex. 5.4.10. Scaling invariance (7.1.4) suggests that this is no coincidence. In the following table, the ideal I_k and its basis are chosen to make the associated form q_k reduced. Moreover, I_k is in the ideal class $[P_2]^k$. The third column lists η_k, the solution of $q_k(x, 1) = 0$ with positive imaginary part. The η_k are pictured in Fig. 7.1.

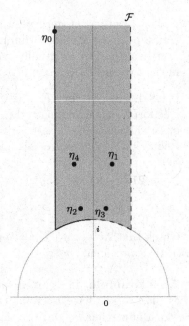

k	$I_k = \mathbb{Z}\alpha + \mathbb{Z}\beta$	$q_k = q_{I_k,\alpha,\beta}$	η_k
0	$\mathbb{Z} + \mathbb{Z}(-\bar{\delta})$	$x^2 + xy + 12y^2$	$\frac{-1+\sqrt{-47}}{2}$
1	$\mathbb{Z} \cdot 2 + \mathbb{Z}\delta$	$2x^2 - xy + 6y^2$	$\frac{1+\sqrt{-47}}{4}$
2	$\mathbb{Z} \cdot 3 + \mathbb{Z}(-\bar{\delta})$	$3x^2 + xy + 4y^2$	$\frac{-1+\sqrt{-47}}{6}$
3	$\mathbb{Z} \cdot 3 + \mathbb{Z}\delta$	$3x^2 - xy + 4y^2$	$\frac{1+\sqrt{-47}}{6}$
4	$\mathbb{Z} \cdot 2 + \mathbb{Z}(-\bar{\delta})$	$2x^2 + xy + 6y^2$	$\frac{-1+\sqrt{-47}}{4}$

Fig. 7.1 Reduced forms of discriminant -47.

The table gives bijections between the following three sets:

(a) the ideal class group $\mathrm{Cl}(\mathbb{Q}[\sqrt{-47}])$;
(b) proper equivalence classes of forms, for which the reduced forms provide a convenient set of representatives;
(c) quadratic numbers of the form $(-b + \sqrt{-47})/2a$ contained in the shaded region $\mathcal{F}$. The interior of $\mathcal{F}$ is $\{z \in \mathbb{C} : \mathrm{Im}\, z > 0, |\mathrm{Re}\, z| < 1/2, |z| > 1\}$.

These bijections are an instance of our main result on forms, Thm. 7.8.1. The correspondence between (a) and (b) gives us a quick way to compute the class number using only simple manipulations with integers, as in Ex. 7.1.9. The bijection between (b) and (c) associates to each form a sort of "ID number", or parameter, in the complex plane. Geometrically, the reduced forms are precisely those whose parameter lies in the region $\mathcal{F}$.   □

As $[P_2]^4 = [P_2]^{-1} = [\bar{P}_2]$, it's not surprising that the ideal $\bar{P}_2$ appears on the last line of the table in Fig. 7.1. But why choose the basis $\{2, -\bar{\delta}\}$ over the more obvious $\{2, \bar{\delta}\}$? What is special about the region $\mathcal{F}$? Why consider proper equivalence of forms, rather than the plain kind? We will answer these (as it turns out, related) questions as we carefully develop the theory of quadratic forms.

Exercises

7.1.1. Verify the formula for $q_{\mathscr{I},\alpha,\beta}(x, y)$ in Ex. 7.1.3. Show that the g.c.d. of the coefficients is 1.

7.1.2.* Check that the ideals I and P_2 of Ex. 7.1.8 are in the same ideal class.

7.1.3. Verify the table in Fig. 7.1.

(a) Check that the ideal I_n is in the ideal class $[P_2]^n$.
(b) Show that the five numbers in the last column of the table are the only quadratic numbers of the form $(-b + \sqrt{-47})/2a$ in the region $\mathcal{F}$.

7.1.4. The region $\mathcal{F}$ in Fig. 7.1 doesn't include the right half of its boundary. How is that related to the second condition of (7.1.7)?

7.1.5. Let $q(x, y) = ax^2 + bxy + cy^2$ be a quadratic form with $a > 0$ and discriminant $b^2 - 4ac < 0$.

(a) Fill in the blanks by completing the square: $ax^2 + bxy + cy^2 = \_(x + \_y)^2 + \_y^2$. Deduce that $q(x, y) \geq (4ac - b^2)/4a$ for $y \neq 0$.
(b) Prove that $q(x, y) \geq (a - |b| + c)\min(x^2, y^2)$. Deduce that $q(x, y) \geq a - |b| + c$ when $x, y \neq 0$.

7.1.6.* We show here that no two of the five forms of Ex. 7.1.9 are properly equivalent.

(a) Show that the forms $2x^2 + xy + 6y^2$ and $2x^2 - xy + 6y^2$ are related by a linear change of variables $\begin{bmatrix} j & k \\ l & m \end{bmatrix} \in \mathrm{GL}_2(\mathbb{Z})$, hence have the same set of values. An analogous statement holds for the forms $3x^2 \pm xy + 4y^2$.
(b) Can we choose the matrix $\begin{bmatrix} j & k \\ l & m \end{bmatrix}$ to have determinant 1?
(c) Show that the three (pairs of) forms from Ex. 7.1.9:
$$x^2 + xy + 12y^2, \quad 2x^2 \pm xy + 6y^2, \quad 3x^2 \pm xy + 4y^2,$$
have different sets of values. Deduce that none can be obtained from the others by a linear change of variables in $\mathrm{GL}_2(\mathbb{Z})$.

7.1.7. For each $D \in \{-23, -26, -71\}$, list all reduced forms $ax^2 + bxy + cy^2$ of discriminant $b^2 - 4ac = D$. Prove that no two are properly equivalent, following Exer. 7.1.6.

7.1.8. Fix a $D \in \mathbb{Z}$. Show that only finitely many forms $ax^2 + bxy + cy^2$ with $b^2 - 4ac = D$ satisfy (7.1.7).

7.2 Elementary Theory of Quadratic Forms

The form $2x^2 - 6xy + 4y^2 = 2(x-y)(x-2y)$ is not interesting: all its values are even because the coefficients are, and we can study it in terms of its linear factors. To exclude such forms, we need some basic definitions. Fix notation by putting $q(x,y) = ax^2 + bxy + cy^2$.

7.2.1 Definition. *A form is **decomposable** if it is a product of two linear polynomials with coefficients in* $\mathbb{Q}$*:* $q(x,y) = a(x - ry)(x - sy)$ *for* $a \in \mathbb{Z}$ *and* $r, s \in \mathbb{Q}$*. Otherwise, the form is **indecomposable**.*

7.2.2 Definition. *To a form* q *we associate two invariants with values in* $\mathbb{Z}$*:*

(a) *the **content**,* $\operatorname{cont} q = \gcd(a,b,c)$*. If* $\operatorname{cont} q = 1$*, the form is called **primitive**.*

(b) *the **discriminant**,* $\operatorname{disc} q = b^2 - 4ac = (a(\eta - \bar{\eta}))^2$*, where* η *is any solution to* $q(x,1) = ax^2 + bx + c = 0$*.*

7.2.3 Definition. *A **quadratic form** is an indecomposable primitive form.*

The form $q(x,y)$ is decomposable precisely when $\operatorname{disc} q$ is a perfect square, as is obvious from the first equality in the following:

$$
\begin{aligned}
ax^2 + bxy + cy^2 &= a\left(x - \tfrac{-b+\sqrt{\operatorname{disc} q}}{2a}y\right)\left(x - \tfrac{-b-\sqrt{\operatorname{disc} q}}{2a}y\right)\\
&= a\left(\left(x + \tfrac{b}{2a}y\right)^2 - \tfrac{\operatorname{disc} q}{4a^2}y^2\right).
\end{aligned}
$$
(7.2.4)

The second equality allows us to determine the possible signs of values of $q(x,y)$ in terms of its discriminant.

7.2.5 Definition. *An indecomposable form* q *is called:*

(a) **indefinite**, *if* $\operatorname{disc} q > 0$*. By (7.2.4),* q *takes both positive and negative values.*

(b) **definite**, *if* $\operatorname{disc} q < 0$*. In that case,* a *and* c *must have the same sign. We further call* q*:*

- **positive definite**, *if* $a, c > 0$*, so* $q(x,y) > 0$ *for all* $x, y \in \mathbb{Z}$*;*
- **negative definite**, *if* $a, c < 0$*, so* $q(x,y) < 0$ *for all* $x, y \in \mathbb{Z}$*.*

First, we give an alternative characterization of content.

7.2.6 Lemma. *The content of a form is the g.c.d. of its values.*

Proof. Let $I = \langle q(r,s) : r,s \in \mathbb{Z}\rangle$ be the ideal of $\mathbb{Z}$ generated by all the values of $q(x,y)$ (see Exer. 2.2.1). We need to show that $I = \mathbb{Z}a + \mathbb{Z}b + \mathbb{Z}c$. One inclusion follows from $q(r,s) = (r^2)a + (rs)b + (s^2)c \in \mathbb{Z}a + \mathbb{Z}b + \mathbb{Z}c$. For the other, observe that the coefficients of q,

$$
a = q(1,0),\ c = q(0,1),\ b = q(1,1) - q(1,0) - q(0,1),
$$
(7.2.7)

are $\mathbb{Z}$-linear combinations of values of $q(x,y)$, and are therefore in I. ∎

7.2.8 Example. Let $\mathscr{I} = \mathbb{Z}\alpha + \mathbb{Z}\beta$ be a fractional ideal in the quadratic field F. We claim that $q_{\mathscr{I},\alpha,\beta}$ of Ex. 7.1.3 is a quadratic form. Without loss of generality, we may assume that $\mathscr{I} = I$, an ideal of $\mathcal{O}$. Indeed, scaling invariance (7.1.4) shows that $q_{\mathscr{I},\alpha,\beta} = q_{n\mathscr{I},n\alpha,n\beta}$ for any $n \in \mathbb{N}$, in particular for those with $n\mathscr{I} \subseteq \mathcal{O}$.

We have a factorization

$$q_{I,\alpha,\beta}(x,y) = \frac{\mathrm{N}(x\alpha - y\beta)}{\mathrm{N}I} = \frac{\mathrm{N}\alpha}{\mathrm{N}I}\left(x - (\beta/\alpha)y\right)\left(x - (\bar{\beta}/\bar{\alpha})y\right).$$

Being a basis of I, α and β are linearly independent over $\mathbb{Q}$. Therefore β/α is irrational, and $q_{I,\alpha,\beta}$ is indecomposable. As $\mathrm{N}I = \gcd(\mathrm{N}\gamma : \gamma \in I)$ (Exer. 4.6.5), we get that

$$\mathrm{cont}\, q_{I,\alpha,\beta} = \gcd(\mathrm{N}(r\alpha - s\beta)/\mathrm{N}I : r,s \in \mathbb{Z}) = 1.$$

The quadratic equation $q_{I,\alpha,\beta}(x,1) = 0$ has β/α as one solution. To compute disc $q_{I,\alpha,\beta}$, we will find another quadratic equation satisfied by β/α, and compare the discriminants of the two. As in Ch. 4, we write the ring of integers of F as $\mathcal{O} = \mathbb{Z}[\delta]$, where $\delta^2 - t\delta + n = 0$, $t,n \in \mathbb{Z}$ and $t^2 - 4n = D_F$. Multiplication by δ is a linear transformation $F \to F$. Let $\left[\begin{smallmatrix} j & k \\ l & m \end{smallmatrix}\right]$ be its matrix with respect to basis $\{\alpha, \beta\}$: $\delta\alpha = j\alpha + l\beta, \delta\beta = k\alpha + m\beta$. Dividing the second equality by the first, we get

$$\beta/\alpha = \frac{k + m(\beta/\alpha)}{j + l(\beta/\alpha)}.$$

Since $\delta I \subseteq I$, the entries j, k, l, m are in $\mathbb{Z}$. By Exer. 4.1.1, $\mathrm{tr}\left[\begin{smallmatrix} j & k \\ l & m \end{smallmatrix}\right] = \mathrm{Tr}\,\delta = t$ and $\det\left[\begin{smallmatrix} j & k \\ l & m \end{smallmatrix}\right] = \mathrm{N}\delta = n$. Thus, β/α satisfies the equation $lx^2 + (j-m)x - k = 0$, with discriminant

$$(j-m)^2 + 4kl = (j+m)^2 - 4(jm - kl) = t^2 - 4n = D_F.$$

We claim that $\gcd(k, j-m, l) = 1$. If p is a prime dividing all three, then $p^2 \mid D_F$. Since 4 is the only square that can divide D_F, we must have $p = 2$; you should check that doesn't happen.

We have two equations with relatively prime integer coefficients, both satisfied by β/α:

$$q_{I,\alpha,\beta}(x,1) = \frac{\mathrm{N}\alpha}{\mathrm{N}I}x^2 - \frac{\mathrm{Tr}(\bar{\alpha}\beta)}{\mathrm{N}I}x + \frac{\mathrm{N}\beta}{\mathrm{N}I} = 0,\ \text{and}$$
$$lx^2 + (j-m)x - k = 0.$$

The equations must be the same up to the sign, so their discriminants are equal: disc $q_{I,\alpha,\beta} = D_F$. $\qquad\square$

Occasionally, we will want to prove that a function on Λ_0 is a form without finding its coefficients. Check that the following condition achieves that, by distinguishing forms among all functions $\Lambda_0 \to \mathbb{Z}$.

7.2.9 Proposition. *A function* $q: \Lambda_0 \to \mathbb{Z}$ *is a form if and only if, for all* $v, w \in \Lambda_0$,

(7.2.10) $$q(v + w) + q(v - w) = 2(q(v) + q(w)).$$

In that case, $q(x, y) = ax^2 + bxy + cy^2$, *with* a, b, c *given by (7.2.7). Conversely, a form* $q(x, y) = ax^2 + bxy + cy^2$, *can be written as a function of a vector* $v \in \Lambda_0$ *by*

$$q(v) = v^t \begin{bmatrix} a & \frac{b}{2} \\ \frac{b}{2} & c \end{bmatrix} v.$$

Exercises

7.2.1. Show that the following three conditions on a form q are equivalent:

(a) q is decomposable.
(b) The equation $q(x, 1) = 0$ has a solution $x \in \mathbb{Q}$.
(c) The equation $q(x, y) = 0$ has a solution $x, y \in \mathbb{Z}$.

7.2.2. Let q be a quadratic form. Show that $\operatorname{disc} q = c^2 D_F$, where $c \in \mathbb{Z}$ and F is the quadratic field containing a solution to $q(x, 1) = 0$.

7.2.3. Prove Prop. 7.2.9. Use induction on x and y to show that any $q: \Lambda_0 \to \mathbb{Z}$ satisfying (7.2.10) is of the form $q(\left[\begin{smallmatrix} x \\ y \end{smallmatrix}\right]) = ax^2 + bxy + cy^2$.

7.2.4.* Let $q: \Lambda_0 \to \mathbb{Z}$ be a function satisfying property (7.2.10).

(a) Show directly from this property (without using the fact that q is a quadratic form) that $q(kv) = k^2 q(v)$ for $k \in \mathbb{Z}$.
(b) Prove that the function $B(v, w) = \frac{1}{2}(q(v+w) - q(v) - q(w))$ is a symmetric bilinear form: $B(v + v', w) = B(v, w) + B(v', w)$ and $B(v, w) = B(w, v)$.
(c) Express $q(v)$ in terms of B.

7.2.5. Given a primitive form q and a prime $p \in \mathbb{N}$, prove that there exist relatively prime $r, s \in \mathbb{Z}$, for which $p \nmid q(r, s)$.

7.2.6.* Complete the proof of $\operatorname{disc} q_{\mathscr{I}, \alpha, \beta} = D_F$ from Ex. 7.2.8 by checking that k, $j - m$ and l can't all be even.

7.2.7. Let $\Lambda_0' = \{\left[\begin{smallmatrix} x \\ y \end{smallmatrix}\right] \in \Lambda_0 : \gcd(x, y) = 1\}$ be the set of **primitive vectors**. Prove that a vector is primitive if and only if it belongs to a basis of Λ_0. Given a quadratic form $q : \Lambda_0 \to \mathbb{Z}$, show that $q(\Lambda_0) = \bigsqcup_{k=0}^{\infty} k^2 q(\Lambda_0')$.

7.2.8. We say that $d \in \mathbb{Z}$ is **represented** by q if $d = q(r, s)$ for some $r, s \in \mathbb{Z}$. If we can find r and s relatively prime, we say that d is **properly represented** by q. Suppose that $q(x, y) = ax^2 + bxy + cy^2$ satisfies $b^2 - 4ac < 0$ and $|b| < a < c$ (in particular, q is positive definite and reduced). Show that the three smallest numbers properly represented by q are $a < c < a - |b| + c$.

Orders

7.2.9. Let $\mathscr{I} = \mathbb{Z}\alpha + \mathbb{Z}\beta \subset F$ be a subgroup with $\mathcal{O}_\Lambda = \mathcal{O}_c$, so that it is a fractional ideal for $\mathcal{O}_c$ and no bigger order. Prove that the form $q_{\mathscr{I}, \alpha, \beta}(x, y) = \mathrm{N}(x\alpha - y\beta)/\mathrm{N}_c\mathscr{I}$ is a quadratic form of discriminant disc $\mathcal{O}_c = c^2 D_F$.

7.3 Parameter of a Quadratic Form

In (7.2.4) we factored a form as $q(x, y) = a(x - \eta_+ y)(x - \eta_- y)$, where $\eta_\pm$ are the roots of the equation $q(x, 1) = ax^2 + bx + c = 0$. We'd like to systematically choose one of $\eta_\pm$ as an "ID number" that uniquely determines q. Since $\eta_\pm$ don't change when we replace q by an integer multiple, it makes sense to consider only primitive forms. If q is decomposable, η_+ and η_- are two unrelated rational numbers, and there is no natural reason to choose one over the other. By contrast, when q is indecomposable, $\eta_\pm$ are conjugate quadratic numbers. Uniformly choosing one of them will establish a one-to-one correspondence between

$$\mathcal{Q} = \text{set of all quadratic forms, and}$$

$$\mathcal{H} = \text{set of all quadratic numbers (real or imaginary).}$$

It's important to remember that a quadratic form is by definition primitive and indecomposable, while a quadratic number is an irrational element of a real or imaginary quadratic field.

Choosing between η_+ and η_- amounts to picking the sign of the square root in the quadratic formula. For any $d \in \mathbb{Z}$, we agree that $\sqrt{d}$ stands for the positive value of the square root when $d > 0$, and the value with positive imaginary part when $d < 0$. Fixing a choice of the square root allows us to define analogs of real and imaginary parts of an element in a quadratic field F.

7.3.1 Definition. *The* **rational** *and the* **irrational** *part of $\alpha \in F$ are given, respectively, by*

$$\mathrm{Ra}\,\alpha = \frac{\alpha + \bar{\alpha}}{2}, \quad \mathrm{Ir}\,\alpha = \frac{\alpha - \bar{\alpha}}{2\sqrt{D_F}}.$$

Note that $\mathrm{Ra}\,\alpha, \mathrm{Ir}\,\alpha \in \mathbb{Q}$, and that $\alpha = \mathrm{Ra}\,\alpha + \mathrm{Ir}\,\alpha\sqrt{D_F}$. Since $\mathcal{H}$ is the disjoint union of all quadratic fields, minus the rationals, the above formulas define functions $\mathrm{Ra}, \mathrm{Ir} : \mathcal{H} \to \mathbb{Q}$. (By the same token, the norm functions on all quadratic fields compile into a single function $\mathrm{N} : \mathcal{H} \to \mathbb{Q}$.) With this terminology, we are ready to define the promised bijection.

7.3.2 Definition. *The* **parameter** *function* $\iota_{\mathcal{Q}\mathcal{H}} : \mathcal{Q} \to \mathcal{H}$ *is given by*

$$q(x,y) = ax^2 + bxy + cy^2 \quad \mapsto \quad \eta_q = \frac{-b + \sqrt{\mathrm{disc}\, q}}{2a},$$

7.3.3 Proposition. *The function* $\iota_{\mathcal{Q}\mathcal{H}}$ *is a bijection. The parameter* η_q *is determined by the following two conditions: (a)* $q(\eta_q, 1) = 0$*, and (b)* $(\mathrm{Ir}\,\eta_q)/a > 0$.

Proof. To construct the inverse, start with $\eta \in \mathcal{H}$ satisfying $e\eta^2 + f\eta + g = 0$ with $e, f, g \in \mathbb{Z}$ and $\gcd(e, f, g) = 1$. If necessary, multiply the equation by -1 to ensure that e has the same sign as $\mathrm{Ir}\,\eta$. Then $\iota_{\mathcal{H}\mathcal{Q}}$ then sends η to the form $q_\eta(x,y) = ex^2 + fxy + gy^2$.

Clearly, η_q satisfies the conditions of the second statement. For the converse, assume that $q(\eta, 1) = 0$ and $(\mathrm{Ir}\,\eta)/a > 0$. By the first condition, $\eta = \eta_q$ or $\bar{\eta}_q$. The latter is impossible, since $(\mathrm{Ir}\,\bar{\eta}_q)/a = -\sqrt{\mathrm{disc}\, q / D_F}/2a^2 < 0$. ∎

We denote the inverse of $\iota_{\mathcal{Q}\mathcal{H}}$ by $\iota_{\mathcal{H}\mathcal{Q}} : \eta \mapsto q_\eta$. This allows us to concisely extend the definition of a discriminant from quadratic integers (Def. 4.2.3) to an arbitrary quadratic numbers.

7.3.4 Definition. *For* $\eta \in \mathcal{H}$*, we put* $\mathrm{disc}\,\eta = \mathrm{disc}\, q_\eta$.

If $a\eta^2 + b\eta + c = 0$ with $a, b, c \in \mathbb{Z}$ relatively prime, it's easy to check that $\mathrm{disc}\,\eta = [a(\eta - \bar{\eta})]^2 = (2a\,\mathrm{Ir}\,\eta)^2 D_F \in \mathbb{Z}$.

7.3.5 Example. A direct computation shows that

$$\iota_{\mathcal{Q}\mathcal{H}}(27x^2 - 95xy + 84y^2) = \frac{95 + \sqrt{-47}}{54},$$

$$\iota_{\mathcal{Q}\mathcal{H}}(-27x^2 + 95xy - 84y^2) = \frac{95 - \sqrt{-47}}{54}.$$

In general, we have $\eta_{-q} = \bar{\eta}_q$. The discriminant of $(95 \pm \sqrt{-47})/54$ is -47. For a general way of reading off the discriminant of a quadratic number, see Exer. 7.3.3. □

Multiplying by -1 turns a positive definite quadratic form into a negative definite one. It thus suffices, for each $D \in \mathbb{Z}, D < 0$, to restrict our attention to the sets

$$\mathcal{Q}_D = \{q \in \mathcal{Q} : \mathrm{disc}\, q = D, a > 0\}, \quad \mathcal{H}_D = \{\eta \in \mathcal{H} : \mathrm{disc}\,\eta = D, \mathrm{Ir}\,\eta > 0\}$$

of positive definite forms of discriminant D and their parameters.

Indefinite forms have no such dichotomy, as they take values of either sign. We therefore simply put, for each $D \in \mathbb{Z}, D > 0$ which is not a perfect square,

$$\mathcal{Q}_D = \{q \in \mathcal{Q} : \operatorname{disc} q = D\}, \quad \mathcal{H}_D = \{\eta \in \mathcal{H} : \operatorname{disc} \eta = D\}.$$

By definition, the parameter function preserves the discriminant, giving the following statement.

7.3.6 Corollary. *For a nonsquare $D \in \mathbb{Z}$, the parameter function $q \mapsto \eta_q$ gives a bijection $\mathcal{Q}_D \to \mathcal{H}_D$, still denoted by $\iota_{\mathcal{Q}\mathcal{H}}$.*

Exercises

7.3.1. Let $\eta \in \mathcal{H}$, and let $q_\eta(x,y) = ax^2 + bxy + cy^2$ be the corresponding quadratic form. Show that a is determined by the following conditions:

(a) $\mathbb{Z}a = \{n \in \mathbb{Z} : n\eta$ is an algebraic integer$\}$, and
(b) $\operatorname{sgn} \operatorname{Ir} \eta = \operatorname{sgn} a$

We think of $a \in \mathbb{Z}$, up to sign, as the smallest denominator of η.

7.3.2. Let $q(x,y) = ax^2 + bxy + cy^2$ have parameter η_q. Verify that $q(x,y) = a\mathrm{N}(x - y\eta_q)$, where N is the norm in the field $\mathbb{Q}[\sqrt{\operatorname{disc} q}]$.

7.3.3. By Exer. 6.5.3, we can write any quadratic number η as $(m + \sqrt{d})/v$, where $m, v, d \in \mathbb{Z}$ and $v \mid m^2 - d$. Write $d = s^2 r$ with r square-free. We may assume that m, v, d are minimal, i.e., $\gcd(m, v, s) = 1$. Under those conditions, show that $\operatorname{disc} \eta = d$ if v is even, and $4d$ if v is odd.

7.3.4.* Let $q(x,y)$ be a quadratic form of discriminant D_F. If $\operatorname{Ir} \eta_q > 0$, prove that $q = q_{\mathbb{Z}+\mathbb{Z}\eta_q, 1, \eta_q}$. What about $\operatorname{Ir} \eta_q < 0$?

7.3.5. Put $-\mathcal{Q}_D = \{-q(x,y) : q(x,y) \in \mathcal{Q}_D\}$ and $\bar{\mathcal{H}}_D = \{\bar{\eta} : \eta \in \mathcal{H}_D\}$. Prove the disjoint union decompositions:

$$\mathcal{Q} = \bigsqcup_{D \in \mathbb{Z}} \mathcal{Q}_D \sqcup \bigsqcup_{D < 0} -\mathcal{Q}_D, \quad \mathcal{H} = \bigsqcup_{D \in \mathbb{Z}} \mathcal{H}_D \sqcup \bigsqcup_{D < 0} \bar{\mathcal{H}}_D.$$

7.3.6. Show that, for $D > 0$ square-free, $\mathbb{Q}[\sqrt{D}] = \mathbb{Q} \sqcup \bigsqcup_{c \in \mathbb{N}} \mathcal{H}_{c^2 D}$. State and prove the analog for $D < 0$.

Orders

7.3.7. Show that $\mathcal{O}_{\mathbb{Z}\alpha + \mathbb{Z}\beta} = \mathcal{O}_c$, for c given by $\operatorname{disc}(\beta/\alpha) = c^2 D_F$. As in Exer. 4.2.13, $\mathcal{O}_{\mathbb{Z}\alpha + \mathbb{Z}\beta}$ is the ring of multipliers of $\mathbb{Z}\alpha + \mathbb{Z}\beta$.

7.4 Linear Symmetries

We next explore what happens to a form and its parameter after a linear change of variables. For our definition, it's best to think of a form as a function on column vectors. Fix, for this section, the notation $A = \begin{bmatrix} j & k \\ l & m \end{bmatrix} \in \operatorname{GL}_2(\mathbb{Z})$.

7.4.1 Definition. *For a form* $q\colon \Lambda_0 \to \mathbb{Z}$, *define the function* $Aq\colon \Lambda_0 \to \mathbb{Z}$ *by*

$$(Aq)(v) = q(A^{-1}v) = q(mx - ky, -lx + jy).$$

The reason for using A^{-1} will become apparent in Ex. 7.5.3. The Definition generalizes the identity (7.1.5), which, in the new notation, reads simply $q_{\mathscr{I}\alpha+l\beta,k\alpha+j\beta} = \begin{bmatrix} j & k \\ l & m \end{bmatrix} q_{\mathscr{I},\alpha,\beta}$. Similarly, the formulas before Prop. 7.1.6 are special cases of the Definition for matrices S and T^k, where $S = \begin{bmatrix} 0 & -1 \\ 1 & 0 \end{bmatrix}$ and $T = \begin{bmatrix} 1 & 1 \\ 0 & 1 \end{bmatrix}$.

7.4.2 Proposition. *Let* $q(x, y)$ *be a form and* $A \in \mathrm{GL}_2(\mathbb{Z})$. *The function* Aq *is also a form: there exist* $a', b', c' \in \mathbb{Z}$ *such that* $(Aq)(x, y) = a'x^2 + b'xy + c'y^2$. *Moreover,* $a' = q(m, -l)$, $c' = q(-k, j)$, *and* $\mathrm{cont}\, Aq = \mathrm{cont}\, q$.

Proof. We check that Aq satisfies the condition of Prop. 7.2.9:

$$\begin{aligned}
(Aq)(v + w) + (Aq)(v - w) &= q(A^{-1}(v + w)) + q(A^{-1}(v - w)) \\
&= q(A^{-1}v + A^{-1}w) + q(A^{-1}v - A^{-1}w) \\
&= 2(q(A^{-1}v) + q(A^{-1}w)) \\
&= 2((Aq)(v) + (Aq)(w)).
\end{aligned}$$

The formulas for the coefficients of Aq follow directly from (7.2.7). Since $A \in \mathrm{GL}_2(\mathbb{Z})$, we have $A^{-1}\Lambda_0 = \Lambda_0$. The forms q and Aq take the same values and therefore have the same content, by Lemma 7.2.6. ∎

7.4.3 Corollary. *If* q *is a quadratic form, so is* Aq.

Proof. The form Aq is primitive, as $\mathrm{cont}\, Aq = \mathrm{cont}\, q = 1$. If Aq were reducible, Exer. 7.2.1 would produce a vector $v \in \Lambda_0$ with $(Aq)(v) = q(A^{-1}v) = 0$. By the same exercise, that would mean that q is reducible; but it is not. ∎

The change of variables in a quadratic form should be reflected in its parameter. Every $\eta \in \mathcal{H}$ is irrational, so $l\eta + m \neq 0$ for all $l, m \in \mathbb{Z}$, and the following definition makes sense.

7.4.4 Definition. *For* $A \in \mathrm{GL}_2(\mathbb{Z})$, *the* **linear fractional transformation** *by* A *is the function* $\mathcal{H} \to \mathcal{H}$ *given by*

$$A\eta = \frac{j\eta + k}{l\eta + m}.$$

7.4.5 Lemma. *For* $A \in \mathrm{GL}_2(\mathbb{Z})$ *and* $\eta \in \mathcal{H}$, $\mathrm{Ir}\, A\eta = \det A \, \mathrm{Ir}\, \eta / \mathrm{N}(l\eta + m)$.

The definition of $A\eta$ was chosen to ensure that A moves η_q "in sync" with the corresponding form q, at least when $\det A = 1$.

7.4.6 Proposition. *Let* $A \in \mathrm{SL}_2(\mathbb{Z})$, *and let* q *be a quadratic form with parameter* η_q. *The parameter of* Aq *is* $\eta_{Aq} = A\eta_q$.

Proof. We check that $A\eta_q$ satisfies the two conditions of Prop. 7.3.3 characterizing η_{Aq}. Condition (a):

$$(Aq)(A\eta_q, 1) = (Aq)\left(\begin{bmatrix} \frac{j\eta_q + k}{l\eta_q + m} \\ 1 \end{bmatrix}\right) = \frac{1}{(l\eta_q + m)^2}(Aq)\left(\begin{bmatrix} j\eta_q + k \\ l\eta_q + m \end{bmatrix}\right)$$

$$= \frac{1}{(l\eta_q + m)^2}(Aq)\left(A\begin{bmatrix} \eta_q \\ 1 \end{bmatrix}\right) = \frac{1}{(l\eta_q + m)^2}q\left(A^{-1}A\begin{bmatrix} \eta_q \\ 1 \end{bmatrix}\right)$$

$$= \frac{1}{(l\eta_q + m)^2}q(\eta_q, 1) = 0.$$

We only use the assumption $\det A = 1$ to verify condition (b). By Prop. 7.4.2 and Exer. 7.3.2, the coefficient of x^2 in $(Aq)(x, y)$ is $a' = q(m, -l) = a\mathrm{N}(l\eta_{\hat{q}} + m)$. Then Lemma 7.4.5 gives

$$\frac{\mathrm{Ir}\, A\eta_q}{a'} = \frac{\frac{\det A}{\mathrm{N}(l\eta_q + m)}\mathrm{Ir}\, \eta_q}{a\mathrm{N}(l\eta_q + m)} = \frac{1}{\mathrm{N}(l\eta_q + m)^2}\frac{\mathrm{Ir}\, \eta_q}{a} > 0. \qquad \blacksquare$$

7.4.7 Corollary. *Let $q \in \mathcal{Q}$, $\eta \in \mathcal{H}$, and $A \in \mathrm{SL}_2(\mathbb{Z})$. Then $\mathrm{disc}\, Aq = \mathrm{disc}\, q$ and $\mathrm{disc}\, A\eta = \mathrm{disc}\, \eta$. The function $q \mapsto Aq$ (resp. $\eta \mapsto A\eta$) sends $\mathcal{Q}_D$ (resp. $\mathcal{H}_D$) to itself.*

Proof. By Def. 7.3.4,

$$\mathrm{disc}\, Aq = (2a'\, \mathrm{Ir}\, A\eta_q)^2 D_F = \left(2a\mathrm{N}(l\eta_q + m)\frac{(\det A)\,\mathrm{Ir}\, \eta_q}{\mathrm{N}(l\eta_q + m)}\right)^2 D_F$$

$$= (2a\,\mathrm{Ir}\, \eta_q)^2 D_F = \mathrm{disc}\, q.$$

Then $\mathrm{disc}\, A\eta = \mathrm{disc}\, q_{A\eta} = \mathrm{disc}\, Aq_\eta = \mathrm{disc}\, q_\eta = \mathrm{disc}\, \eta$. When $D > 0$, this proves that A preserves both $\mathcal{Q}_D$ and $\mathcal{H}_D$.

When $D < 0$, we need to check the additional positivity conditions. The form $(Aq)(x, y) = a'x^2 + b'xy + c'y^2$ is positive definite. Indeed, $a' = q(m, -l) > 0$, being a value of the positive definite form $q \in \mathcal{Q}_D$. On the parameter side, we again use the assumption $\det A = 1$. If $\eta \in \mathcal{H}_D$, then $\mathrm{Ir}\, A\eta = (\det A)\,\mathrm{Ir}\, \eta/\mathrm{N}(l\eta + m) > 0$, since the norm of an element in an imaginary quadratic field is always positive. $\qquad \blacksquare$

Exercises

7.4.1. Let $A = \begin{bmatrix} j & k \\ l & m \end{bmatrix} \in \mathrm{GL}_2(\mathbb{Z})$ and $q(x, y) = ax^2 + bxy + cy^2$. Check by explicit computation that

$$(Aq)(x, y) = (am^2 - bml + cl^2) - (2amk - bmj - bkl + 2clj)$$
$$+ (ak^2 - bkj + cj^2).$$

7.4.2. Let $T = \left[\begin{smallmatrix} 1 & 1 \\ 0 & 1 \end{smallmatrix}\right], S = \left[\begin{smallmatrix} 0 & -1 \\ 1 & 0 \end{smallmatrix}\right]$. Given a form q, show that the formulas before Prop. 7.1.6 compute $T^k q$ and Sq, respectively.

7.4.3. Let $q(x, y)$ be a quadratic form, and fix $d \in \mathbb{Z}$. Prove that there exists an $A \in \mathrm{GL}_2(\mathbb{Z})$ for which $(Aq)(x, y) = ax^2 + bxy + cy^2$ satisfies $\gcd(a, d) = 1$.

7.4.4.* In this exercise, Aq and $A\eta$ have the same meaning as in the text for $A \in \mathrm{SL}_2(\mathbb{Z})$. By slightly changing the definition of one or the other for $A \in \mathrm{GL}_2(\mathbb{Z}) \setminus \mathrm{SL}_2(\mathbb{Z})$, we can ensure that the identity $\eta_{Aq} = A\eta_q$ remains valid.

(a) Do this by keeping the definition of Aq, and putting

$$\left[\begin{smallmatrix} j & k \\ l & m \end{smallmatrix}\right]\eta = \frac{j\bar{\eta} + k}{l\bar{\eta} + m}, \quad \text{when } jm - kl = -1.$$

(b) Keeping Def. 7.4.4 as the definition of $A\eta$ for all $A \in \mathrm{GL}_2(\mathbb{Z})$, find the right re-definition of Aq for $A \in \mathrm{GL}_2(\mathbb{Z}) \setminus \mathrm{SL}_2(\mathbb{Z})$.

7.5 Group Actions

In the preceding section we studied how *one* matrix changes a quadratic form and its parameter. We compile the effect of *all* the matrices in $\mathrm{GL}_2(\mathbb{Z})$ or $\mathrm{SL}_2(\mathbb{Z})$ into a single structure with the help of the following definition.

7.5.1 Definition. *Let G be a group and X a set. A (left)* **group action** *of G on X (a G-action on X, for short) is a function $G \times X \to X$, denoted $(g, x) \mapsto gx$, that satisfies the following axioms for all $g, g' \in G$ and $x \in X$:*

(a) $1_G x = x$, and
(b) $(gg')x = g(g'x)$.

For a fixed $g \in G$, $x \mapsto gx$ is a bijection $X \to X$. We think of it as g permuting the elements of X in a manner compatible with the group operation on G (see Exer. 7.5.4 for a formalization of this point of view).

7.5.2 Example. Let $H \subseteq G$ be a subgroup and $Y \subseteq X$ a subset preserved by H: if $h \in H$ and $y \in Y$, then $hy \in Y$ also. The group action $G \times X \to X$ restricts to a function $H \times Y \to Y$ which also satisfies the group action axioms. We say that we **restricted** the action of G on X to an action of H on Y. $\square$

For the next two examples, take $A \in \mathrm{GL}_2(\mathbb{Z})$, a quadratic form $q \in \mathcal{Q}$, and a quadratic number $\eta \in \mathcal{H}$.

7.5.3 Example. By Cor. 7.4.3, Aq is also in $\mathcal{Q}$. The function $(A, q) \mapsto Aq$ satisfies the axioms of a group action with $G = \mathrm{GL}_2(\mathbb{Z}), X = \mathcal{Q}$:

(a) $(\left[\begin{smallmatrix} 1 & 0 \\ 0 & 1 \end{smallmatrix}\right]q)(v) = q(\left[\begin{smallmatrix} 1 & 0 \\ 0 & 1 \end{smallmatrix}\right]^{-1} v) = q(v)$

(b) $[(AB)q](v) = q((AB)^{-1}v) = q(B^{-1}A^{-1}v) = (Bq)(A^{-1}v) = [A(Bq)](v)$

The A^{-1} in the definition of the action was essential for A and B to appear in the correct order in the last expression. By Cor. 7.4.7, this action restricts to an action $\mathrm{GL}_2(\mathbb{Z}) \times \mathcal{Q}_D \to \mathcal{Q}_D$, and further to an action of $\mathrm{SL}_2(\mathbb{Z})$ on $\mathcal{Q}_D$, for any nonsquare $D \in \mathbb{Z}$. $\square$

7.5.4 Example. It's easy to check that the formula $(A, \eta) \mapsto A\eta \in \mathcal{H}$ defines an action of $\mathrm{GL}_2(\mathbb{Z})$ on $\mathcal{H}$. For $D < 0$, this action does *not* restrict to an action of $\mathrm{GL}_2(\mathbb{Z})$ on $\mathcal{H}_D$: if $A \in \mathrm{GL}_2(\mathbb{Z}) \setminus \mathrm{SL}_2(\mathbb{Z})$ and $\eta \in \mathcal{H}_D$, Lemma 7.4.5 gives Ir $A < 0$, so that $A\eta \notin \mathcal{H}_D$. The same lemma shows that $(A, \eta) \mapsto A\eta$ does restrict to an action of $\mathrm{SL}_2(\mathbb{Z})$ on $\mathcal{H}_D$. $\square$

Group actions give rise to interesting equivalence relations.

7.5.5 Definition. *Let $x, y \in X$. If $y = gx$ for some $g \in G$, we write $x \sim y$ and say that x and y are G-**equivalent** . The **orbit** of x is the set $Gx = \{gx : g \in G\}$.*

The use of the term "equivalent" is justified by the following proposition, the proof off which is a good exercise.

7.5.6 Proposition. *The relation $x \sim y$ is an equivalence relation on X. The equivalence class of x is Gx. The following three statements are equivalent: (a) $x \sim y$; (b) $y \in Gx$; (c) $Gx = Gy$.*

In traditional terminology introduced in Ex. 7.1.3, two forms are equivalent (resp. properly equivalent) precisely when they are $\mathrm{GL}_2(\mathbb{Z})$- (resp. $\mathrm{SL}_2(\mathbb{Z})$-) equivalent.

7.5.7 Definition. *The **quotient** of X by G is the set of all orbits,*

$$X/G = \{Gx : x \in X\}.$$

*A **fundamental domain** for the action of G on X is a subset $\mathcal{F} \subset X$ such that for each $x \in X$ there exists a unique $y \in \mathcal{F}$ equivalent to x.*

Sending an element in X to its orbit defines a surjection $X \to X/G$, which restricts to a bijection $\mathcal{F} \xrightarrow{\sim} X/G$. You can think of a fundamental domain as a concrete model for the quotient X/G, obtained by picking a preferred representative in each equivalence class.

7.5.8 Example. Recall the matrices $S = \begin{bmatrix} 0 & -1 \\ 1 & 0 \end{bmatrix}$ and $T = \begin{bmatrix} 1 & 1 \\ 0 & 1 \end{bmatrix}$ from Sec. 7.4. The Definite Reduction Algorithm of Prop. 7.1.6 produces a sequence of matrices $A_1, \ldots, A_r$, each equal to S or a power of T, such that $(A_1 \ldots A_r)q$ has small coefficients. As S and T are in $\mathrm{SL}_2(\mathbb{Z})$, this shows that any form q is properly equivalent to a form $ax^2 + bxy + cy^2$ with $|b| \leq |a| \leq |c|$, and $b > 0$ if $|a|$ equals $|b|$ or $|c|$. There are five such forms of discriminant -47, as we found in Ex. 7.1.9:

$$\mathcal{F} = \{x^2 + xy + 12y^2, 2x^2 \pm xy + 6y^2, 3x^2 \pm xy + 4y^2\}.$$

Exercise 7.1.6 shows that no two of these forms are properly equivalent. In other words, $\mathcal{F}$ is a fundamental domain for the action of $\mathrm{SL}_2(\mathbb{Z})$ on $\mathcal{Q}_{-47}$. $\qquad\square$

When studying a structure on a set, we invariably study structure-preserving functions as well: groups and homomorphisms, topological spaces and continuous functions, sets with a G-action and equivariant functions.

7.5.9 Definition. *Let $\varphi\colon X \to Y$ be a function between two sets, each with a G-action. We say that φ is G-equivariant if, for all $g \in G$ and $x \in X$,*

$$\varphi(gx) = g\varphi(x).$$

On the left, g acts on X, and on the right, it acts on Y.

7.5.10 Proposition. *Let a group G act on two sets X and Y, and let $\varphi\colon X \to Y$ be a G-equivariant function. The expression $\tilde\varphi(Gx) = G\varphi(x)$ gives a well-defined function $\tilde\varphi\colon X/G \to Y/G$. We say that $\tilde\varphi$ is **induced** from φ by **passing to the quotient**.*

Proof. If $Gx = Gy$, then there exists a $g \in G$ with $y = gx$. By the G-equivariance of φ, this implies $\varphi(y) = \varphi(gx) = g\varphi(x)$, so that $G\varphi(x) = G\varphi(y)$ by Prop. 7.5.6. Thus, $\tilde\varphi(Gx)$ depends only on the orbit Gx and not on the particular choice of representative. $\qquad\blacksquare$

7.5.11 Example. With the new terminology, Prop. 7.4.6 says that the parameter function $\iota_{\mathcal{Q}\mathcal{H}}\colon \mathcal{Q} \to \mathcal{H}$ is an $\mathrm{SL}_2(\mathbb{Z})$-equivariant bijection. By Prop. 7.5.10, it induces a bijection of sets $\mathcal{Q}/\mathrm{SL}_2(\mathbb{Z}) \cong \mathcal{H}/\mathrm{SL}_2(\mathbb{Z})$, and further

$$\mathcal{Q}_D/\mathrm{SL}_2(\mathbb{Z}) \cong \mathcal{H}_D/\mathrm{SL}_2(\mathbb{Z}).$$

Our goal in the rest of the chapter is to further identify these two sets with (a version of) the ideal class group, at least when D is the discriminant of a quadratic field. $\qquad\square$

We finish with a dry but simple lemma that will be crucial in Sec. 7.7; you can skip it until then. It concerns two groups, G and H, both acting on a set X. To distinguish the two actions, we denote them by $(g, x) \mapsto g \cdot x$ for $g \in G$, and $(h, x) \mapsto h * x$ for $h \in H$. The orbits of x are denoted $G \cdot x$ and $H * x$, respectively.

The two actions **commute** if for all $g \in G, h \in H$, and $x \in X$,

$$g \cdot (h * x) = h * (g \cdot x).$$

Here's a useful way of reading this equation: the function $x \mapsto g \cdot x$ is equivariant with respect to the action of H. It induces, by Prop. 7.5.10, a function $g\colon X/H \to X/H$. This is easily checked to define an action of G on X/H, explicitly given by $g \cdot (H * x) = H * (g \cdot x)$. By symmetry, we also get an action of H on X/G, and can therefore form two quotients $(X/H)/G$ and $(X/G)/H$. The upshot is that the order in which we take the quotients doesn't matter.

7.5.12 Lemma. *The assignment $G \cdot (H * x) \mapsto H * (G \cdot x)$ gives a well-defined bijection $(X/H)/G \to (X/G)/H$.*

Proof. We need to show that if $G \cdot (H * x) = G \cdot (H * y)$, then $H * (G \cdot x) = H * (G \cdot y)$. Once we have that, we get bijectivity for free: the same argument with the roles of G and H reversed shows that the inverse is well-defined.

We have the following chain of implications:

$$G \cdot (H * x) = G \cdot (H * y) \Rightarrow$$
$$g \cdot (H * x) = H * y, \text{ for some } g \in G \Rightarrow$$
$$H * (g \cdot x) = H * y \Rightarrow$$
$$h * (g \cdot x) = y, \text{ for some } h \in H.$$

Since the two actions commute, this means that $g \cdot (h * x) = y$, hence $G \cdot y = G \cdot (h * x) = h * (G \cdot x)$, and finally $H * (G \cdot x) = H * (G \cdot y)$. ∎

Exercises

7.5.1. Here are two examples of group actions:

(a) Let K be a field. Check that the usual matrix multiplication of a 2×2 matrix and a 2×1 column vector defines an action of $GL_2(K)$ on K^2. What is the quotient of this action?

(b) Show that a subgroup $H \subseteq G$ acts on the group G by left translations, $(h, g) \mapsto hg$. If G is abelian, prove that the quotient of this action is G/H, the usual set of cosets of H. What if G isn't abelian?

7.5.2. Fix a group G. For the given set X, check that the indicated function $G \times X \to X$ is an action of G on X.

(a) any X; $(g, x) \mapsto x$ (the **trivial** action)

(b) $X = G$; $(g, x) \mapsto gxg^{-1}$ (the **conjugation** action)

(c) $X = G/H = \{xH : x \in G\}$, for a subgroup $H \subseteq G$, not necessarily normal; $(g, xH) \mapsto gxH$. (First check that this is independent of the choice of coset representative x.)

7.5.3. Let G act on X, and let $\varphi \colon H \to G$ be a group homomorphism. Prove that $(h, x) \mapsto \varphi(h)x$ defines an action of H on X.

7.5.4. Given an action of G on X, the function $\tau_g : x \mapsto gx$ belongs to $\text{Bij}(X)$, the group of all bijections from X to itself. Show that $g \mapsto \tau_g$ is a group homomorphism $G \to \text{Bij}(X)$. Conversely, starting with such a homomorphism, reconstruct the action $G \times X \to X$.

7.5.5. Let $G = \{z \in \mathbb{C} : |z| = 1\}$ act on $\mathbb{C}$ by $(e^{i\varphi}, z) \mapsto e^{i\varphi}z$. In other words, $e^{i\varphi}$ acts as the rotation of the plane with angle φ. Describe all orbits of this action, and draw a few of them. The picture should suggest the origin of the term "orbit."

7.5.6. Let X be a set with a G-action, and let $x \in X$. The **stabilizer** of x in G is defined as $G_x = \{g \in G : gx = x\}$.

(a) Show that G_x is a subgroup of G.
(b) Given $h \in G$, what is the relationship between G_x and G_{hx}?
(c) Show that $hG_x \mapsto hx$ gives a well-defined G-equivariant bijection $G/G_x \to Gx \subseteq X$. Here $G/G_x = \{gG_x : g \in G\}$.

7.5.7. Find a fundamental domain for the action of $\mathrm{GL}_2(\mathbb{Z})$ on $\mathcal{Q}_{-47}$.

7.5.8. Verify that your re-definition of $A\eta$ in Exer. 7.4.4 (a) defines a group action of $\mathrm{GL}_2(\mathbb{Z})$ on $\mathcal{H}$.

7.5.9. Let X be a set with a G-action, Y an arbitrary set, and $f: X \to Y$ a function satisfying $f(gx) = f(x)$ for all $g \in G, x \in X$. Show that $\tilde{f}(Gx) = f(x)$ gives a well-defined function $\tilde{f}: X/G \to Y$. Do this directly, or apply Prop. 7.5.10 with the trivial G-action on Y, as in Exer. 7.5.2 (a).

7.5.10. A **right group action** of G on X is a function $X \times G \to X$, denoted $(x, g) \mapsto xg$, which satisfies the following axioms for all $g, g' \in G, x \in X$: (a) $x1_G = x$, and (b) $x(gg') = (xg)g'$. Right and left actions differ in more than just typography. When gg' acts on the right, the first element, g, acts first; when gg' acts on the left, g acts second.

(a) If $(g, x) \mapsto gx$ is a left action, show that $(x, g) \mapsto g^{-1}x$ is a right action.
(b) If G acts on X on the left, construct a natural action of G on the set of functions $\mathscr{F}(X) = \{f : X \to \mathbb{C}\}$. Is it a left or a right action?

7.5.11. The set X has two commuting actions by groups G and H.

(a) Show that $G \times H$ acts on X by $(g, h)x = g \cdot (h * x)$.
(b) Since $(g, 1_H)$ and $(1_G, h)$ commute in $G \times H$, it is essential to assume that their actions on X commute. Where did you use that assumption in the proof of (a)?
(c) Re-prove Lemma 7.5.12 by showing that $(X/G)/H \cong X/(G \times H) \cong (X/H)/G$.

7.6 Orientation

In Sec. 7.1 we hinted at a correspondence between ideal classes of an imaginary quadratic field F and reduced quadratic forms of discriminant D_F. To construct it, let's first try to send the ideal class $[\mathscr{I}]$ to the orbit $\mathrm{GL}_2(\mathbb{Z})q_{\mathscr{I},\alpha,\beta}$, where (α, β) is any ordered basis of $\mathscr{I}$. This assignment is well-defined: changing the basis of $\mathscr{I}$ changes $q_{\mathscr{I},\alpha,\beta}$ to a $\mathrm{GL}_2(\mathbb{Z})$-equivalent form, and choosing a different representative of $[\mathscr{I}]$ doesn't change the form either, by (7.1.4). The resulting function $\varphi : \mathrm{Cl}(F) \to \mathcal{Q}_{D_F}/\mathrm{GL}_2(\mathbb{Z})$ is unfortunately not a bijection, as it can't distinguish between $[\mathscr{I}]$ and $[\bar{\mathscr{I}}]$:

$$q_{\bar{\mathscr{I}},\bar{\alpha},\bar{\beta}}(x, y) = \frac{\mathrm{N}(x\bar{\alpha} - y\bar{\beta})}{\mathrm{N}\bar{\mathscr{I}}} = \frac{\mathrm{N}(x\alpha - y\beta)}{\mathrm{N}\mathscr{I}} = q_{\mathscr{I},\alpha,\beta}(x, y).$$

To get the correct correspondence between ideals and forms, and to include real quadratic fields, we impose an orientation condition on the basis (α, β). From now on, F is an arbitrary quadratic field.

Any $z, w \in \mathbb{C}$ span a parallelogram in the complex plane. Up to sign, the area of this parallelogram is $\operatorname{Im} \bar{z}w = (\bar{z}w - z\bar{w})/2i$. This expression is positive precisely when the angle from z to w is oriented counterclockwise. This inspires the following definition.

7.6.1 Definition. *For* $\alpha, \beta \in F$, *put*

$$\langle \alpha, \beta \rangle = \frac{\bar{\alpha}\beta - \alpha\bar{\beta}}{2\sqrt{D_F}} = \operatorname{N}\alpha \operatorname{Ir}(\beta/\alpha).$$

This pairing satisfies $\langle \bar{\alpha}, \bar{\beta} \rangle = -\langle \alpha, \beta \rangle = \langle \beta, \alpha \rangle$. The value of $\sqrt{D_F}$ is fixed in the usual way, as in Sec. 7.3. It is evident from the second expression in the Definition that $\langle \alpha, \beta \rangle \in \mathbb{Q}$. Even when F is imaginary, the following definition thus makes sense.

7.6.2 Definition. *An ordered pair* (α, β) *of elements of F is* **oriented** *if* $\langle \alpha, \beta \rangle > 0$. *An* **oriented fractional ideal** *is a triple* $(\mathscr{I}, \alpha, \beta)$, *where $\mathscr{I}$ is a fractional ideal in F with an oriented basis* (α, β).

7.6.3 Example. Let $I = \mathbb{Z}a + \mathbb{Z}\alpha$ with $a \mid \operatorname{N}\alpha$ be an ideal of $\mathcal{O}$ in standard form. The triple (I, a, α) is an oriented fractional ideal if and only if $\langle a, \alpha \rangle = a \operatorname{Ir} \alpha > 0$. $\qquad\qquad \square$

The set of all oriented fractional ideals in F,

$$\mathcal{I}_F = \{(\mathscr{I}, \alpha, \beta) : \mathscr{I} = \mathbb{Z}\alpha + \mathbb{Z}\beta \text{ fractional ideal in } F, \langle \alpha, \beta \rangle > 0\},$$

has a rich structure.

Conjugation: We would like to associate to an oriented fractional ideal $(\mathscr{I}, \alpha, \beta)$ a "conjugate" of the form $(\bar{\mathscr{I}}, ?, ?)$. Our guess from the beginning of this section, $(\bar{\mathscr{I}}, \bar{\alpha}, \bar{\beta})$, doesn't work since $(\bar{\alpha}, \bar{\beta})$ is not oriented: $\langle \bar{\alpha}, \bar{\beta} \rangle = -\langle \alpha, \beta \rangle < 0$. This suggests defining the conjugate by

$$\overline{(\mathscr{I}, \alpha, \beta)} = (\bar{\mathscr{I}}, \bar{\beta}, \bar{\alpha}),$$

with associated quadratic form

$$q_{\bar{\mathscr{I}}, \bar{\beta}, \bar{\alpha}}(x, y) = \frac{\operatorname{N}(x\bar{\beta} - y\bar{\alpha})}{\operatorname{N}\bar{\mathscr{I}}} = \frac{\operatorname{N}(x\beta - y\alpha)}{\operatorname{N}\mathscr{I}} = q_{\mathscr{I}, \alpha, \beta}(y, x).$$

In other words, $q_{\bar{\mathscr{I}}, \bar{\beta}, \bar{\alpha}}$ and $q_{\mathscr{I}, \alpha, \beta}$ are related by $\left[\begin{smallmatrix} 0 & 1 \\ 1 & 0 \end{smallmatrix}\right] \in \operatorname{GL}_2(\mathbb{Z}) \setminus \operatorname{SL}_2(\mathbb{Z})$. They may also be related by a matrix in $\operatorname{SL}_2(\mathbb{Z})$, but apart from such special cases we will see that $\operatorname{SL}_2(\mathbb{Z})q_{\mathscr{I}, \alpha, \beta} \neq \operatorname{SL}_2(\mathbb{Z})q_{\bar{\mathscr{I}}, \bar{\beta}, \bar{\alpha}}$. Proper equivalence can distinguish between an oriented ideal and its conjugate, which partly explains our interest in $\operatorname{SL}_2(\mathbb{Z})$-actions.

7.6.4 Example. Let's go back to the field $F = \mathbb{Q}[\sqrt{-47}]$, the ideal $P_2 = \mathbb{Z} \cdot 2 + \mathbb{Z}\delta$, and the corresponding form $q_{P_2,2,\delta}(x,y) = 2x^2 - xy + 6y^2$. The form associated to the conjugate ideal, $q_{\bar{P}_2,\bar{\delta},2}(x,y) = 6x^2 - xy + 2y^2$, reduces to $2x^2 + xy + 6y^2$. In general, conjugating an ideal corresponds, up to proper equivalence, to changing the sign of the xy-coefficient of the associated form. $\square$

7.6.5 Example. Even though $q_{\mathscr{I},\bar{\beta},\bar{\alpha}}$ and $q_{\mathscr{I},\alpha,\beta}$ are always related by $\left[\begin{smallmatrix}0 & 1\\ 1 & 0\end{smallmatrix}\right] \notin \mathrm{SL}_2(\mathbb{Z})$, they can, in some instances, also be related by a matrix in $\mathrm{SL}_2(\mathbb{Z})$. Consider the forms

$$q_{I,17,7+\sqrt{15}}(x,y) = 17x^2 - 14xy + 2y^2,$$
$$q_{\bar{I},7-\sqrt{15},17}(x,y) = 2x^2 - 14xy + 17y^2.$$

They correspond to conjugate oriented ideals in $\mathbb{Q}[\sqrt{15}]$, but are nevertheless in the same $\mathrm{SL}_2(\mathbb{Z})$-orbit, as $q_{\bar{I},7-\sqrt{15},17} = (T^7 S)q_{I,17,7+\sqrt{15}}$. This reflects the fact that the class of $I = \mathbb{Z} \cdot 17 + \mathbb{Z}(7 + \sqrt{15})$ is of order 2 in $\mathrm{Cl}(\mathbb{Q}[\sqrt{15}])$. $\square$

Scaling: For $\gamma \in F$ and $(\mathscr{I}, \alpha, \beta) \in \mathcal{I}_F$ put $\gamma(\mathscr{I}, \alpha, \beta) = (\gamma\mathscr{I}, \gamma\alpha, \gamma\beta)$. As $\langle \gamma\alpha, \gamma\beta \rangle = (\mathrm{N}\gamma)\langle \alpha, \beta \rangle$, this is an oriented ideal if and only if

$$\gamma \in F_+^\times = \{\gamma \in F : \mathrm{N}\gamma > 0\}.$$

The set $\mathcal{I}_F$ thus has an action of $F_+^\times$, but not of the larger group $F^\times$.

$\mathrm{SL}_2(\mathbb{Z})$-action: For $\left[\begin{smallmatrix}j & k\\ l & m\end{smallmatrix}\right] \in \mathrm{SL}_2(\mathbb{Z})$ and $(\mathscr{I}, \alpha, \beta) \in \mathcal{I}_F$, put

$$(7.6.6) \qquad \left[\begin{smallmatrix}j & k\\ l & m\end{smallmatrix}\right] (\mathscr{I}, \alpha, \beta) = (\mathscr{I}, m\alpha + l\beta, k\alpha + j\beta).$$

Check directly that this formula defines a group action. To see it conceptually, observe that matrix multiplication

$$(7.6.7) \qquad \left[\begin{smallmatrix}j & k\\ l & m\end{smallmatrix}\right]\left[\begin{smallmatrix}\alpha\\ \beta\end{smallmatrix}\right] = \left[\begin{smallmatrix}j\alpha+k\beta\\ l\alpha+m\beta\end{smallmatrix}\right]$$

defines an action of $\mathrm{SL}_2(\mathbb{Z})$ on column vectors with entries in F. Writing that action in typographical disguise, and restricting to oriented pairs (α, β) for which $\mathscr{I} = \mathbb{Z}\alpha + \mathbb{Z}\beta$ is a fractional ideal, we get an action on $\mathcal{I}_F$:

$$\left[\begin{smallmatrix}j & k\\ l & m\end{smallmatrix}\right] * (\mathscr{I}, \alpha, \beta) = (\mathscr{I}, j\alpha + k\beta, l\alpha + m\beta).$$

We only need to check that the above triple is still an oriented fractional ideal. This follows from Exer. 7.6.1, since $\left[\begin{smallmatrix}j & k\\ l & m\end{smallmatrix}\right]$ has determinant 1. Here it is crucial that we're restricting to an action of $\mathrm{SL}_2(\mathbb{Z})$. With the aid of the conjugation homomorphism

$$\chi\left(\left[\begin{smallmatrix}j & k\\ l & m\end{smallmatrix}\right]\right) = \left[\begin{smallmatrix}0 & 1\\ 1 & 0\end{smallmatrix}\right]\left[\begin{smallmatrix}j & k\\ l & m\end{smallmatrix}\right]\left[\begin{smallmatrix}0 & 1\\ 1 & 0\end{smallmatrix}\right]^{-1} = \left[\begin{smallmatrix}m & l\\ k & j\end{smallmatrix}\right],$$

formula (7.6.6) can be written as $A(\mathscr{I}, \alpha, \beta) = \chi(A) * (\mathscr{I}, \alpha, \beta)$. By Exer. 7.5.3, it defines a group action. The "twist" by χ will be necessary to ensure compatibility of actions in Thm. 7.8.2.

Exercises

7.6.1. For $\alpha, \beta \in F$ and $\left[\begin{smallmatrix} j & k \\ l & m \end{smallmatrix}\right] \in \mathrm{GL}_2(\mathbb{Z})$, show that $\langle j\alpha + k\beta, l\alpha + m\beta \rangle = (\det \left[\begin{smallmatrix} j & k \\ l & m \end{smallmatrix}\right]) \langle \alpha, \beta \rangle$.

7.6.2. Apply Prop. 4.1.3 (d) to the first expression in Def. 7.6.1 to check that $\langle \alpha, \beta \rangle \in \mathbb{Q}$.

7.6.3. Let $I = \mathbb{Z}\alpha + \mathbb{Z}\beta$ be an ideal of $\mathcal{O}$. Prove that $\mathrm{N}I = 2 \, |\langle \alpha, \beta \rangle|$.

7.6.4. Show that scaling by $\alpha \in F_+^{\times}$ commutes with action of $A \in \mathrm{SL}_2(\mathbb{Z})$ on $\mathcal{I}_F$. Which of those two actions commute with the conjugation?

Orders

7.6.5. Fix a conductor c. Define the set of oriented invertible fractional ideals of $\mathcal{O}_c$ by $\mathcal{I}_{\mathcal{O}_c} = \{(\mathscr{I}, \alpha, \beta) : \mathscr{I} = \mathbb{Z}\alpha + \mathbb{Z}\beta, \mathcal{O}_{\mathscr{I}} = \mathcal{O}_c, \langle \alpha, \beta \rangle > 0\}$. Prove the analogs of the results on $\mathcal{I}_F$ in this section for $\mathcal{I}_{\mathcal{O}_c}$

7.7 The Narrow Ideal Class Group

The actions of $F_+^{\times}$ and $\mathrm{SL}_2(\mathbb{Z})$ on $\mathcal{I}_F$ commute. Setting $X = \mathcal{I}_F$, $G = F_+^{\times}$ and $H = \mathrm{SL}_2(\mathbb{Z})$ in Lemma 7.5.12 gives a natural identification

$$(\mathcal{I}_F/F_+^{\times})/\mathrm{SL}_2(\mathbb{Z}) \cong (\mathcal{I}_F/\mathrm{SL}_2(\mathbb{Z}))/F_+^{\times}.$$

The left side will play a key role in the next section; here we study the right side. First, the inside quotient:

7.7.1 Proposition. *Sending the orbit* $\mathrm{SL}_2(\mathbb{Z})(\mathscr{I}, \alpha, \beta)$ *to* $\mathscr{I}$ *defines a bijection* $\tilde{\varphi} : \mathcal{I}_F/\mathrm{SL}_2(\mathbb{Z}) \xrightarrow{\sim} \mathbb{I}_F$.

Proof. Define a function $\varphi : \mathcal{I}_F \to \mathbb{I}_F$ by $\varphi(\mathscr{I}, \alpha, \beta) = \mathscr{I}$. This function is $\mathrm{SL}_2(\mathbb{Z})$-invariant:

$$\varphi\left(\left[\begin{smallmatrix} j & k \\ l & m \end{smallmatrix}\right](\mathscr{I}, \alpha, \beta)\right) = \varphi(\mathscr{I}, m\alpha + l\beta, k\alpha + j\beta) = \mathscr{I} = \varphi(\mathscr{I}, \alpha, \beta).$$

By Exer. 7.5.9, φ induces a well-defined function $\mathcal{I}_F/\mathrm{SL}_2(\mathbb{Z}) \to \mathbb{I}_F$ which is none other than $\tilde{\varphi}$. As $\tilde{\varphi}$ is clearly onto, we only need to check that it is injective. Assume $\tilde{\varphi}(\mathrm{SL}_2(\mathbb{Z})(\mathscr{I}, \alpha, \beta)) = \tilde{\varphi}(\mathrm{SL}_2(\mathbb{Z})(\mathscr{I}', \alpha', \beta'))$, so that

$\mathscr{I} = \mathscr{I}'$. Then (α, β) and (α', β'), being two bases of $\mathscr{I}$, are related by a matrix $A \in \mathrm{GL}_2(\mathbb{Z})$. By Exer. 7.6.1, $\langle \alpha', \beta' \rangle = (\det A)\langle \alpha, \beta \rangle$. Since both bases are oriented, $\det A$ must be positive, so $\det A = 1$ and $A \in \mathrm{SL}_2(\mathbb{Z})$. As $A(\mathscr{I}, \alpha, \beta) = (\mathscr{I}', \alpha', \beta')$, the basic properties of orbits from Prop. 7.5.6 give $\mathrm{SL}_2(\mathbb{Z})(\mathscr{I}, \alpha, \beta) = \mathrm{SL}_2(\mathbb{Z})(\mathscr{I}', \alpha', \beta')$. ∎

The bijection φ is equivariant for the scaling action of $F_+^\times$, and induces a bijection $(\mathcal{I}_F / \mathrm{SL}_2(\mathbb{Z}))/F_+^\times \cong \mathbb{I}_F/F_+^\times$. The latter is the quotient of $\mathbb{I}_F$ by the equivalence relation which identifies $\mathscr{I}$ and $\mathscr{J}$ precisely when $\mathscr{J} = \alpha\mathscr{I}$ for some $\alpha \in F_+^\times$. As $\alpha\mathscr{I} = (\mathcal{O}\alpha)\mathscr{I}$, this is just the equivalence relation defining the group quotient of $\mathbb{I}_F$ by the subgroup $\mathbb{P}_F^+$, of the following definition.

7.7.2 Definition. *Let F be a quadratic field.*

(a) *A principal ideal is* **totally positive** *if has the form $\mathcal{O}\alpha$ with $\mathrm{N}\alpha > 0$. Such ideals form a subgroup $\mathbb{P}_F^+ \subseteq \mathbb{P}_F \subseteq \mathbb{I}_F$.*
(b) *The* **narrow ideal class group** *of F is the quotient group $\mathrm{Cl}^+(F) = \mathbb{I}_F/\mathbb{P}_F^+$.*
(c) *The* **narrow class number** *of F is $h^+(F) = |\mathrm{Cl}^+(F)|$.*

The narrow ideal class group is clearly a variation on the ideal class group. It's "narrow" because we take the quotient by a smaller subgroup. Implicit in the definition of $h^+(F)$ is the fact that $\mathrm{Cl}^+(F)$ is finite. For F imaginary quadratic, that is obvious, as $F_+^\times = F^\times$ and therefore $\mathrm{Cl}^+(F) = \mathrm{Cl}(F)$. For real quadratic fields, we will see in Prop. 7.7.5 that $\mathrm{Cl}^+(F)$ is either equal to $\mathrm{Cl}(F)$, or double its size.

7.7.3 Example. Let $F = \mathbb{Q}[\sqrt{2}]$ and $\mathcal{O} = \mathbb{Z}[\sqrt{2}]$. Is the principal ideal $\mathcal{O}\sqrt{2}$ totally positive? Since $\mathrm{N}\sqrt{2} = -2$, you might be tempted to say no. But think again: $\mathcal{O}\sqrt{2}$ is in $\mathbb{P}_F^+$ precisely when it has *some* generator in $F_+^\times$. Such a generator is $\sqrt{2}(1+\sqrt{2})$: as $1+\sqrt{2}$ is a unit of norm -1, $\mathrm{N}(\sqrt{2}(1+\sqrt{2})) = 2$. The same argument shows that $\mathbb{P}_F^+ = \mathbb{P}_F$, and therefore $\mathrm{Cl}^+(F) = \mathrm{Cl}(F)$, whenever $\mathcal{O}^\times$ contains a unit of norm -1. □

7.7.4 Example. Let $F = \mathbb{Q}[\sqrt{15}]$ and $\mathcal{O} = \mathbb{Z}[\sqrt{15}]$. Any generator of $\mathcal{O}(3 + \sqrt{15})$ is of the form $(3 + \sqrt{15})\varepsilon$ for some unit ε. For our ideal to be totally positive, we'd need $\mathrm{N}\varepsilon = -1$, since $\mathrm{N}(3 + \sqrt{15}) = -6$. But there is no such $\varepsilon = x + y\sqrt{15}$: reducing $x^2 - 15y^2 = \mathrm{N}\varepsilon = -1$ modulo 3 produces a contradiction. The ideal $\mathcal{O}(3 + \sqrt{15})$ is principal, but not totally positive. The class of $\mathcal{O}(3+\sqrt{15})$ in $\mathrm{Cl}^+(F)$ is not the identity, even though its class in $\mathrm{Cl}(F)$ is. □

You will have no problem generalizing these examples to prove the following proposition.

7.7.5 Proposition. *A well-defined, surjective group homomorphism $\psi : \mathrm{Cl}^+(F) \to \mathrm{Cl}(F)$ is given by*

$$\psi(\mathbb{P}_F^+ \cdot \mathscr{I}) = \mathbb{P}_F \cdot \mathscr{I}.$$

It is an isomorphism, unless F is a real quadratic field with a fundamental unit that satisfies $N\varepsilon_F = 1$. In that case, $\ker\psi$ is cyclic of order 2, generated by the coset $\mathbb{P}_F^+ \cdot (\mathcal{O}\sqrt{D})$.

The following combination of results from this section is all we will need in proving the main theorem of the chapter, Thm. 7.8.1

7.7.6 Corollary. *Denote the $F_+^\times$-orbit of $(\mathscr{I}, \alpha, \beta)$ by $[\mathscr{I}, \alpha, \beta]$. The assignment $\mathrm{SL}_2(\mathbb{Z})[\mathscr{I}, \alpha, \beta] \mapsto \mathbb{P}_F^+ \cdot \mathscr{I}$ is a well-defined bijection*

$$(\mathcal{I}_F/F_+^\times)/\mathrm{SL}_2(\mathbb{Z}) \xrightarrow{\sim} \mathrm{Cl}^+(F).$$

Exercises

7.7.1. We know that $[\mathscr{I}]^{-1} = [\bar{\mathscr{I}}]$ in $\mathrm{Cl}(F)$. Is that still true in $\mathrm{Cl}^+(F)$?

7.7.2. If F is a real quadratic field, show that $\mathbb{P}_F^+ = \{\mathcal{O}\alpha : \alpha > 0 \text{ and } \bar{\alpha} > 0\}$. This characterization of $\mathbb{P}_F^+$ generalizes to bigger number fields.

7.7.3. Prove Prop. 7.7.5.

7.7.4.* Let F be a real quadratic field. We say that an ideal class in $\mathrm{Cl}(F)$ or $\mathrm{Cl}^+(F)$ is ramified if it contains a ramified prime ideal.

(a) Show that $\mathrm{Cl}^+(F)[2]$ is generated by ramified classes.
(b) When $N\varepsilon_F = -1$, conclude that $\mathrm{Cl}(F)[2]$ is generated by ramified classes.
(c) Assume that there is a ramified prime $p \in \mathbb{N}$ with $p \equiv 3 \pmod 4$, or $p = 2$. Prove that $\mathrm{Cl}(F)[2]$ is still generated by ramified classes.
(d) For an example where $\mathrm{Cl}(F)[2]$ is no longer generated by ramified classes, we need a field F in which $N\varepsilon_F = 1$, and in which all ramified primes are congruent to 1 mod 4. Moreover, by Exer. 6.7.5 (c), D_F must have at least two prime divisors. Check that these conditions hold for $D_F = 205$ (which is, in fact, the smallest example). Compute $\mathrm{Cl}(\mathbb{Q}[\sqrt{205}])$ to show that its 2-torsion is not generated by ramified classes.

Orders

7.7.5. Fix a conductor c. Resuming the notation from the Orders exercises of Sec. 5.1, we put

$$\mathbb{P}_{1,+}(c) = \{\mathcal{O}(\alpha/\beta) : \alpha, \beta \in \mathcal{O}_c, \gcd(N\alpha, c) = \gcd(N\beta, c) = 1, N(\alpha/\beta) > 0\}.$$

Define the **narrow ideal class group** of $\mathcal{O}_c$ by $\mathrm{Cl}^+(\mathcal{O}_c) = \mathbb{I}_1(c)/\mathbb{P}_{1,+}(c)$. Show that the natural homomorphism $\mathrm{Cl}^+(\mathcal{O}_c) \to \mathrm{Cl}^+(F)$ is surjective, with kernel isomorphic to $[(\mathcal{O}/c\mathcal{O})^\times/(\mathbb{Z}/c\mathbb{Z})^\times]/\varphi(\mathcal{O}_+^\times)$. Here $\mathcal{O}_+^\times = \{\gamma \in \mathcal{O}^\times : N\gamma = 1\}$.

7.8 The Three Avatars

Given a quadratic field F, we write $\mathcal{Q}_F$ for $\mathcal{Q}_{D_F}$, and $\mathcal{H}_F$ for $\mathcal{H}_{D_F}$. Our main result on quadratic forms is an extension of Ex. 7.5.11, where we showed that $\mathcal{Q}_F / \operatorname{SL}_2(\mathbb{Z}) \cong \mathcal{H}_F / \operatorname{SL}_2(\mathbb{Z})$.

7.8.1 Theorem. *There exist bijections*

$$\operatorname{Cl}^+(F) \cong \mathcal{Q}_F / \operatorname{SL}_2(\mathbb{Z}) \cong \mathcal{H}_F / \operatorname{SL}_2(\mathbb{Z}).$$

Proof. By Cor. 7.7.6, $\operatorname{Cl}^+(F) \cong (\mathcal{I}_F / F_+^\times) / \operatorname{SL}_2(\mathbb{Z})$ is a quotient set of an $\operatorname{SL}_2(\mathbb{Z})$-action, like the other two terms. In Thm. 7.8.2 we will construct $\operatorname{SL}_2(\mathbb{Z})$-equivariant bijections $\mathcal{I}_F / F_+^\times \cong \mathcal{Q}_F \cong \mathcal{H}_F$, the latter coming from Cor. 7.3.6. By Prop. 7.5.10 these induce the desired bijections on the quotients by $\operatorname{SL}_2(\mathbb{Z})$. ∎

Theorem 7.8.1 is beautiful mathematics. It lets us study the same object from three different viewpoints:

(a) Since $\mathcal{Q}_F / \operatorname{SL}_2(\mathbb{Z})$ is identified with $\operatorname{Cl}^+(F)$, it too should have a group structure. It is far from obvious how to "compose" two classes of quadratic forms. Legendre almost figured it out, a century before the definition of the ideal class group, but was thrown off by considering $\operatorname{GL}_2(\mathbb{Z})$-, rather than $\operatorname{SL}_2(\mathbb{Z})$-equivalence. It was Gauss who realized the importance of the latter (see Exer. 7.8.4 for a modern explanation of this confusion). The subject has surprising vigor in the twenty-first century, with Bhargava opening an important new perspective on the composition of quadratic forms, which we take up in Sec. 7.11.

(b) It is no more difficult to consider quadratic forms of the more general discriminant $D = c^2 D_F$. We then naturally ask what group, analogous to $\operatorname{Cl}^+(F)$, is in bijection with $\mathcal{Q}_D / \operatorname{SL}_2(\mathbb{Z})$. The answer, the narrow ideal class group of the order $\mathcal{O}_c$ in F, has been studied in the exercises.

(c) We can think of $\mathcal{H}_F \subset F$ as a subset of a plane, either directly when F is imaginary, or via the embedding $\rho : F \hookrightarrow \mathbb{R}^2$ when F is real. We then view $\mathcal{H}_F / \operatorname{SL}_2(\mathbb{Z})$ as the geometric avatar of the narrow ideal class group. Geometry will give us a nice description of the fundamental domain for the action of $\operatorname{SL}_2(\mathbb{Z})$ on $\mathcal{H}_F$, explaining reduction of quadratic forms and allowing us to quickly compute class numbers.

To construct the bijections $\mathcal{I}_F / F_+^\times \cong \mathcal{Q}_F \cong \mathcal{H}_F$, consider the functions

$$\iota_{\mathcal{IH}} : \mathcal{I}_F \to \mathcal{H}_F, \quad (\mathcal{I}, \alpha, \beta) \mapsto \frac{\beta}{\alpha}$$

$$\iota_{\mathcal{IQ}} : \mathcal{I}_F \to \mathcal{Q}_F, \quad (\mathcal{I}, \alpha, \beta) \mapsto q_{\mathcal{I}, \alpha, \beta}(x, y) = \frac{\mathrm{N}(x\alpha - y\beta)}{\mathrm{N}\mathcal{I}}$$

It is clear that $\iota_{\mathcal{I}\mathcal{H}}$ is invariant under the scaling action of $F_+^\times$. Formula (7.1.4) shows that the same is true for $\iota_{\mathcal{I}\mathcal{Q}}$. The two functions they induce on $\mathcal{I}_F/F_+^\times$ are still denoted $\iota_{\mathcal{I}\mathcal{H}}$ and $\iota_{\mathcal{I}\mathcal{Q}}$.

7.8.2 Theorem. *The functions* $\iota_{\mathcal{I}\mathcal{H}} : \mathcal{I}_F/F_+^\times \to \mathcal{H}_F$ *and* $\iota_{\mathcal{I}\mathcal{Q}} : \mathcal{I}_F/F_+^\times \to \mathcal{Q}_F$ *are bijections, with inverses denoted* $\iota_{\mathcal{H}\mathcal{I}}$ *and* $\iota_{\mathcal{Q}\mathcal{I}}$. *These four functions are compatible with the parameter function* $\iota_{\mathcal{Q}\mathcal{H}} : \mathcal{Q}_F \to \mathcal{H}_F$ *of Cor. 7.3.6 and its inverse* $\iota_{\mathcal{H}\mathcal{Q}}$, *in the sense of the composition identities*

$$(7.8.3) \qquad \iota_{\mathcal{H}\mathcal{Q}} \circ \iota_{\mathcal{I}\mathcal{H}} = \iota_{\mathcal{I}\mathcal{Q}}, \quad \iota_{\mathcal{Q}\mathcal{I}} \circ \iota_{\mathcal{H}\mathcal{Q}} = \iota_{\mathcal{H}\mathcal{I}}, \quad \iota_{\mathcal{I}\mathcal{H}} \circ \iota_{\mathcal{Q}\mathcal{I}} = \iota_{\mathcal{Q}\mathcal{H}},$$

and similarly for the inverses:

$$(7.8.4) \qquad \iota_{\mathcal{H}\mathcal{I}} \circ \iota_{\mathcal{Q}\mathcal{H}} = \iota_{\mathcal{Q}\mathcal{I}}, \quad \iota_{\mathcal{Q}\mathcal{H}} \circ \iota_{\mathcal{I}\mathcal{Q}} = \iota_{\mathcal{I}\mathcal{H}}, \quad \iota_{\mathcal{I}\mathcal{Q}} \circ \iota_{\mathcal{H}\mathcal{I}} = \iota_{\mathcal{H}\mathcal{Q}}.$$

Finally, all six bijections are $\mathrm{SL}_2(\mathbb{Z})$-*equivariant.*

Beneath the crowded notation, the theorem simply identifies three sets of very different objects: oriented ideals up to scaling, quadratic forms, and their parameters. Those three sets are the vertices in the directed graph (7.8.5). The solid arrows represent the functions we already have; the two dashed arrows stand for the functions we need to construct. The three loops stand for the identity functions on their respective sets:

(7.8.5)

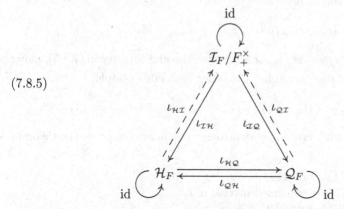

We neatly summarize Thm. 7.8.2 by saying that this diagram is **commutative**: pick any two vertices and any two paths connecting them, and the compositions of the functions along the two paths will be equal. Here are two examples of deducing a composition identity in Thm. 7.8.2 from the commutativity of the diagram (7.8.5):

(a) Consider the following two paths from $\mathcal{I}_F/F_+^\times$ to $\mathcal{Q}_F$:

$$\mathcal{I}_F/F_+^\times \xrightarrow{\iota_{\mathcal{I}\mathcal{H}}} \mathcal{H}_F \xrightarrow{\iota_{\mathcal{H}\mathcal{Q}}} \mathcal{Q}_F \text{ and } \mathcal{I}_F/F_+^\times \xrightarrow{\iota_{\mathcal{I}\mathcal{Q}}} \mathcal{Q}_F.$$

The commutativity of the diagram asserts that the composition of functions along either of these paths is the same: $\iota_{\mathcal{HQ}} \circ \iota_{\mathcal{IH}} = \iota_{\mathcal{IQ}}$. This is the first identity in (7.8.3).

(b) The equality of the compositions along the paths

$$\mathcal{H}_F \xrightarrow{\iota_{\mathcal{HI}}} \mathcal{I}_F/F_+^\times \xrightarrow{\iota_{\mathcal{IH}}} \mathcal{H}_F \text{ and } \mathcal{H}_F \xrightarrow{\mathrm{id}} \mathcal{H}_F$$

gives $\iota_{\mathcal{HI}} \circ \iota_{\mathcal{IH}} = \mathrm{id}_{\mathcal{H}_F}$, so $\iota_{\mathcal{HI}}$ is the left inverse of $\iota_{\mathcal{IH}}$. The other half of the bijectivity of $\iota_{\mathcal{IH}}$ is proved similarly, using the two paths starting and ending at $\mathcal{I}_F/F_+^\times$.

Proof of Thm. 7.8.2. Once we've constructed $\iota_{\mathcal{HI}}$ and $\iota_{\mathcal{QI}}$, we need to prove ten composition identities: the two that show $\iota_{\mathcal{HI}}$ is the two-sided inverse of $\iota_{\mathcal{IH}}$, the two analogous ones for $\iota_{\mathcal{QI}}$, and finally the six listed in (7.8.3) and (7.8.4). We start by removing redundant identities from this list.

We may drop the three identities in (7.8.4), since any one of them arises by inverting the corresponding identity in (7.8.3). For example, if we assume the first identity in (7.8.3), we get

$$\iota_{\mathcal{QI}} = \iota_{\mathcal{IQ}}^{-1} = \left(\iota_{\mathcal{HQ}} \circ \iota_{\mathcal{IH}} \right)^{-1} = \iota_{\mathcal{IH}}^{-1} \circ \iota_{\mathcal{HQ}}^{-1} = \iota_{\mathcal{HI}} \circ \iota_{\mathcal{QH}}.$$

In particular, if $\iota_{\mathcal{HI}}$ exists, we *must* put $\iota_{\mathcal{QI}} = \iota_{\mathcal{HI}} \circ \iota_{\mathcal{QH}}$. The function thus defined is indeed the inverse of $\iota_{\mathcal{IQ}}$:

$$\iota_{\mathcal{IQ}} \circ \iota_{\mathcal{QI}} = \left(\iota_{\mathcal{HQ}} \circ \iota_{\mathcal{IH}} \right) \circ \left(\iota_{\mathcal{HI}} \circ \iota_{\mathcal{QH}} \right) = \iota_{\mathcal{HQ}} \circ \mathrm{id}_{\mathcal{H}_F} \circ \iota_{\mathcal{QH}} = \mathrm{id}_{\mathcal{Q}_F},$$

and similarly $\iota_{\mathcal{QI}} \circ \iota_{\mathcal{IQ}} = \mathrm{id}_{\mathcal{I}_F/F_+^\times}$. Moreover, the first identity in (7.8.3), along with the existence of $\iota_{\mathcal{HI}}$, implies the other two. For example,

$$\iota_{\mathcal{QI}} \circ \iota_{\mathcal{HQ}} = \left(\iota_{\mathcal{HI}} \circ \iota_{\mathcal{QH}} \right) \circ \iota_{\mathcal{HQ}} = \iota_{\mathcal{HI}} \circ \left(\iota_{\mathcal{QH}} \circ \iota_{\mathcal{HQ}} \right) = \iota_{\mathcal{HI}}.$$

This leaves us with just three identities. Specifically, to prove the theorem it is enough to:

(a) construct the function $\iota_{\mathcal{HI}}$;
(b) check that $\iota_{\mathcal{HI}}$ is a two-sided inverse of $\iota_{\mathcal{IH}}$;
(c) prove the first identity of (7.8.3), $\iota_{\mathcal{HQ}} \circ \iota_{\mathcal{IH}} = \iota_{\mathcal{IQ}}$.

Construction of $\iota_{\mathcal{HI}}$: Denote the $F_+^\times$-orbit of $(\mathscr{I}, \alpha, \beta)$ by $[\mathscr{I}, \alpha, \beta]$. For $\eta \in \mathcal{H}_F$, put

$$\iota_{\mathcal{HI}}(\eta) = \begin{cases} [\mathbb{Z} + \mathbb{Z}\eta, 1, \eta] & \text{if } \mathrm{Ir}\,\eta > 0 \\ [\sqrt{D_F}(\mathbb{Z} + \mathbb{Z}\eta), \sqrt{D_F}, \sqrt{D_F}\eta] & \text{if } \mathrm{Ir}\,\eta < 0. \end{cases}$$

The triples inside the brackets make sense as oriented ideals. By Exer. 7.3.7, $\mathbb{Z} + \mathbb{Z}\eta$ is a fractional ideal in F since $\mathrm{disc}\,\eta = D_F$. In the first case, the indicated ordered basis is oriented since $\langle 1, \eta \rangle = \mathrm{Ir}\,\eta > 0$. The second case occurs only for F real, when we have $\langle \sqrt{D_F}, \sqrt{D_F}\eta \rangle = (N\sqrt{D_F})\langle 1, \eta \rangle = -D_F$ Ir $\eta > 0$. The oriented ideals $(\mathbb{Z} + \mathbb{Z}\eta, 1, \eta)$ and $(\sqrt{D_F}(\mathbb{Z} + \mathbb{Z}\eta), \sqrt{D_F}, \sqrt{D_F}\eta)$ need not be $F_+^\times$-equivalent, as the obvious constant of proportionality $\sqrt{D_F}$ is not in $F_+^\times$.

Proof of $\iota_{\mathcal{HI}} = \iota_{\mathcal{IH}}^{-1}$: Recall that $\iota_{\mathcal{IH}}([\mathscr{I}, \alpha, \beta]) = \beta/\alpha$, so that $\iota_{\mathcal{IH}}(\iota_{\mathcal{HI}}(\eta)) = \eta$. To check that $\iota_{\mathcal{HI}}$ is the two-sided inverse of $\iota_{\mathcal{IH}}$, all we need to show is that $\iota_{\mathcal{HI}}(\iota_{\mathcal{IH}}([\mathscr{I}, \alpha, \beta])) = \iota_{\mathcal{HI}}(\beta/\alpha) = [\mathscr{I}, \alpha, \beta]$.

Since $(\mathscr{I}, \alpha, \beta)$ is oriented, $N\alpha$ and Ir β/α have the same sign. If $N\alpha > 0$, then $1/\alpha \in F_+^\times$. This justifies the last equality in

$$\iota_{\mathcal{HI}}\left(\frac{\beta}{\alpha}\right) = \left[\mathbb{Z} + \mathbb{Z}\frac{\beta}{\alpha}, 1, \frac{\beta}{\alpha}\right] = \left[\frac{1}{\alpha}(\mathbb{Z}\alpha + \mathbb{Z}\beta, \alpha, \beta)\right] = [\mathscr{I}, \alpha, \beta].$$

If $N\alpha < 0$, then $\sqrt{D_F}/\alpha \in F_+^\times$, and we get

$$\iota_{\mathcal{HI}}\left(\frac{\beta}{\alpha}\right) = \left[\sqrt{D_F}\left(\mathbb{Z} + \mathbb{Z}\frac{\beta}{\alpha}\right), \sqrt{D_F}, \frac{\sqrt{D_F}\beta}{\alpha}\right]$$

$$= \left[\frac{\sqrt{D_F}}{\alpha}(\mathbb{Z}\alpha + \mathbb{Z}\beta, \alpha, \beta)\right] = [\mathscr{I}, \alpha, \beta].$$

Thus, $\iota_{\mathcal{HI}}$ is the desired inverse of $\iota_{\mathcal{IH}}$.

The identity $\iota_{\mathcal{HQ}} \circ \iota_{\mathcal{IH}} = \iota_{\mathcal{IQ}}$: By the definition of $\iota_{\mathcal{HQ}}$ in Prop. 7.3.3, we need to show that $\iota_{\mathcal{IH}}([\mathscr{I}, \alpha, \beta]) = \beta/\alpha$ is the parameter of the quadratic form $\iota_{\mathcal{IQ}}([\mathscr{I}, \alpha, \beta]) = q_{\mathscr{I}, \alpha, \beta}$. We do this by checking the conditions of Prop. 7.3.3, starting with condition (a):

$$q_{\mathscr{I}, \alpha, \beta}\left(\frac{\beta}{\alpha}, 1\right) = \frac{N\left(\frac{\beta}{\alpha}\alpha - \beta\right)}{N\mathscr{I}} = 0.$$

For condition (b), we saw in Ex. 7.1.3 that the coefficient of x^2 in $q_{\mathscr{I}, \alpha, \beta}(x, y)$ is $N\alpha/N\mathscr{I}$, implying

$$\frac{\text{Ir}(\beta/\alpha)}{N\alpha/N\mathscr{I}} = \frac{N\mathscr{I}}{(N\alpha)^2}\left(N\alpha \text{ Ir }\frac{\beta}{\alpha}\right) = \frac{N\mathscr{I}}{(N\alpha)^2}\langle \alpha, \beta \rangle > 0.$$

Finally, we prove the $SL_2(\mathbb{Z})$-equivariance of the six bijections. In Prop. 7.4.6 we showed it for $\iota_{\mathcal{HQ}}$ and $\iota_{\mathcal{QH}}$. By the composition identities (7.8.3), it is enough to check that $\iota_{\mathcal{IH}}$ is also $SL_2(\mathbb{Z})$-equivariant. The $SL_2(\mathbb{Z})$-action on $\mathcal{I}_F$, defined by (7.6.6), induces an action on $\mathcal{I}_F/F_+^\times$. It was chosen to make the following computation work:

$$\iota_{\mathcal{IH}}\left(\begin{bmatrix} j & k \\ l & m \end{bmatrix}[\mathscr{I}, \alpha, \beta]\right) = \iota_{\mathcal{IH}}([\mathscr{I}, m\alpha + l\beta, k\alpha + j\beta])$$

$$= \frac{k\alpha + j\beta}{m\alpha + l\beta} = \frac{j(\beta/\alpha) + k}{l(\beta/\alpha) + m}$$

$$= \begin{bmatrix} j & k \\ l & m \end{bmatrix} \frac{\beta}{\alpha} = \begin{bmatrix} j & k \\ l & m \end{bmatrix} \iota_{\mathcal{I}\mathcal{H}}([\mathscr{I}, \alpha, \beta]).$$

This proves Thm. 7.8.2, and with it the identifications $\mathrm{Cl}^+(F) \cong \mathcal{Q}_F / \mathrm{SL}_2(\mathbb{Z}) \cong \mathcal{H}_F / \mathrm{SL}_2(\mathbb{Z})$. ∎

7.8.6 Example. Returning to $\mathbb{Q}[\sqrt{-47}]$, the three objects in the left triangle correspond under the bijections of the commutative diagram (7.8.5):

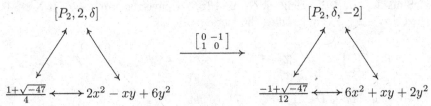

$$[P_2, 2, \delta] \qquad\qquad\qquad\qquad [P_2, \delta, -2]$$

$$\begin{bmatrix} 0 & -1 \\ 1 & 0 \end{bmatrix}$$

$$\frac{1+\sqrt{-47}}{4} \longleftrightarrow 2x^2 - xy + 6y^2 \qquad \frac{-1+\sqrt{-47}}{12} \longleftrightarrow 6x^2 + xy + 2y^2$$

Acting on each of the three objects by $\begin{bmatrix} 0 & -1 \\ 1 & 0 \end{bmatrix} \in \mathrm{SL}_2(\mathbb{Z})$ produces the vertices of the right-hand triangle. By $\mathrm{SL}_2(\mathbb{Z})$-equivariance, they too should be identified by the bijections of Thm. 7.8.2. You are invited to check that directly.

Exercises

7.8.1. Check that the functions $\iota_{\mathcal{I}\mathcal{H}}$ and $\iota_{\mathcal{I}\mathcal{Q}}$ really take values in $\mathcal{H}_F$ and $\mathcal{Q}_F$, respectively.

7.8.2. The vertices in the following graphs stand for sets, and arrows for functions between them. Translate the commutativity of these diagrams into composition identities:

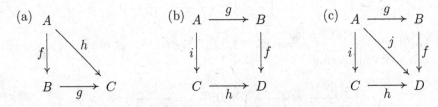

Observe that the commutativity of diagram (b) is equivalent to the commutativity of diagram (c): a commutative square (and any other, more complicated commutative diagram) breaks up into commutative triangles.

7.8.3. Show that if the two small squares $AA'BB'$ and $BB'CC'$ in the diagram below are commutative, then so is the big square $AA'CC'$:

$$\begin{array}{ccccc} A & \xrightarrow{\alpha} & B & \xrightarrow{\beta} & C \\ \downarrow a & & \downarrow b & & \downarrow c \\ A' & \xrightarrow{\alpha'} & B' & \xrightarrow{\beta'} & C' \end{array}$$

7.8.4. Find an equivalence relation $\sim$ on $\mathrm{Cl}^+(F)$ for which $(\mathrm{Cl}^+(F)/\sim) \cong$ $\mathcal{Q}_F/\operatorname{GL}_2(\mathbb{Z})$. Show that the latter quotient doesn't inherit a group structure from $\mathrm{Cl}^+(F)$.

7.8.5. Show that $[q] \in \mathcal{Q}_F/\operatorname{SL}_2(\mathbb{Z})$ corresponds to the identity $[\mathcal{O}] \in \mathrm{Cl}^+(F)$ if and only if the equation $q(x,y) = 1$ has a solution.

7.8.6. If the class $[\mathscr{I}] \in \mathrm{Cl}^+(F)$ corresponds to the $\operatorname{SL}_2(\mathbb{Z})$-orbit of $ax^2 + bxy + cy^2$ under the identification $\mathrm{Cl}^+(F) \cong \mathcal{Q}_F/\operatorname{SL}_2(\mathbb{Z})$, show that $[\mathscr{I}]^{-1}$ corresponds to the orbit of $ax^2 - bxy + cy^2$, which is also the orbit of $cx^2 + bxy + ay^2$.

Orders

7.8.7. For the order $\mathcal{O}_c$ in F of conductor c, prove that $\mathrm{Cl}^+(\mathcal{O}_c) \cong$ $\mathcal{Q}_{c^2 D_F}/\operatorname{SL}_2(\mathbb{Z}) \cong \mathcal{H}_{c^2 D_F}/\operatorname{SL}_2(\mathbb{Z})$.

7.9 Reduced Positive Definite Forms

Let F be an imaginary quadratic field. To visualise $\mathrm{Cl}(F) \cong \mathcal{H}_F/\operatorname{SL}_2(\mathbb{Z})$, we find a fundamental domain for the action of $\operatorname{SL}_2(\mathbb{Z})$ on $\mathcal{H}_F$. We can do this for any discriminant $D < 0$, not just D_F, by viewing $\mathcal{H}_D \subset F$ as a subset of the complex plane. In fact, $\mathcal{H}_D$ is contained in the complex upper half-plane $\mathbb{H} = \{z \in \mathbb{C} : \operatorname{Im} z > 0\}$: by definition, every $\eta \in \mathcal{H}_D$ has $\operatorname{Ir} \eta = \operatorname{Im} \eta / |D_{\mathbb{Q}[\sqrt{D}]}|^{1/2} > 0$. The action of $\operatorname{SL}_2(\mathbb{Z})$ on $\mathcal{H}_D$ by linear fractional transformations extends to an action on all of $\mathbb{H}$, still given by $\left[\begin{smallmatrix} j & k \\ l & m \end{smallmatrix}\right] z = (jz + k)/(lz + m)$. All we need to check is that $\left[\begin{smallmatrix} j & k \\ l & m \end{smallmatrix}\right] z \in \mathbb{H}$ for $\left[\begin{smallmatrix} j & k \\ l & m \end{smallmatrix}\right] \in \operatorname{SL}_2(\mathbb{Z})$ and $z \in \mathbb{H}$:

$$(7.9.1) \qquad \operatorname{Im} \left[\begin{smallmatrix} j & k \\ l & m \end{smallmatrix}\right] z = \frac{jk - lm}{|lz + m|^2} \operatorname{Im} z > 0.$$

Consider the region $\mathcal{F}_{\mathrm{im}}$ in the upper half-plane given by

$$\mathcal{F}_{\mathrm{im}} = \{z \in \mathbb{H} : |\operatorname{Re} z| < \tfrac{1}{2}, |z| > 1\} \sqcup \{-\tfrac{1}{2} + yi : y \geq \tfrac{\sqrt{3}}{2}\}$$
$$\sqcup \{z \in \mathbb{H} : -\tfrac{1}{2} \leq \operatorname{Re} z \leq 0, |z| = 1\},$$

and depicted as the shaded region in Fig. 7.2. The set $\mathcal{F}_{\mathrm{im}}$ includes only the left half of its boundary, drawn in solid lines.

7.9.2 Proposition. *The region $\mathcal{F}_{\mathrm{im}}$ is a fundamental domain for the action of $\operatorname{SL}_2(\mathbb{Z})$ on $\mathbb{H}$.*

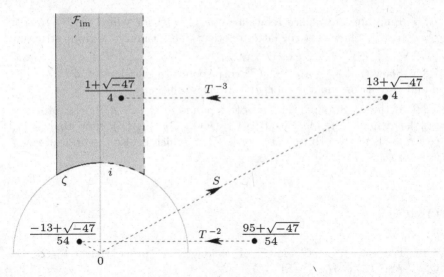

Fig. 7.2 Reduction of $(95 + \sqrt{-47})/54$ to $\mathcal{F}_{\text{im}}$,the fundamental domain for the action of $SL_2(\mathbb{Z})$ on $\mathbb{H}$

Proof. It suffices to show that any $z \in \mathbb{H}$ can be moved to a unique point in $\mathcal{F}_{\text{im}}$ by repeated action of the matrices $S = \left[\begin{smallmatrix} 0 & -1 \\ 1 & 0 \end{smallmatrix}\right]$ and $T = \left[\begin{smallmatrix} 1 & 1 \\ 0 & 1 \end{smallmatrix}\right]$ in $SL_2(\mathbb{Z})$. As $Tz = z + 1$, a suitable power of T will move z to the strip $|\operatorname{Re} z| \leq 1/2$. Since $|Sz| = 1/|z|$, S interchanges the inside and the outside of the half-disc $|z| \leq 1$. Alternating these two transformations should eventually move z to a point simultaneously inside the strip and outside the disc, i.e., in $\mathcal{F}_{\text{im}}$. That is the idea behind the following:

REDUCTION ALGORITHM FOR $\mathbb{H}$

1. INPUT: $z \in \mathbb{H}$. Put $z_0 := z, M := \left[\begin{smallmatrix} 1 & 0 \\ 0 & 1 \end{smallmatrix}\right]$, and set counter $i := 0$.

2. Choose $a \in \mathbb{Z}$ so that $z_i + a$ is in the strip $|w| \leq 1/2$. (To wit, $-a$ is the integer nearest $\operatorname{Re} z_i$.) Put $z_{i+1} := z_i + a = T^a z_i$, $M := T^a M$, and $i := i + 1$.

3. If $|z_i| \geq 1$, go to Step 5.

4. Put $z_{i+1} := -1/z_i = S z_i$, $M := SM$, $i := i + 1$, and go to Step 2.

5. (Boundary condition) If $\operatorname{Re} z_i = 1/2$, put $z_i := z_i - 1$ and $M := T^{-1}M$, so that now $\operatorname{Re} z_i = -1/2$.

6. (Boundary condition) If $|z_i| = 1$ and $0 < \operatorname{Re} z_i < 1/2$, put $z_i := -1/z_i$ and $M := SM$, so that now $|z_i| = 1$ and $-1/2 < \operatorname{Re} z_i < 0$.

7. OUTPUT: $z_i = Mz \in \mathcal{F}_{\text{im}}$ as the point $SL_2(\mathbb{Z})$-equivalent to z, and M as the matrix that moves z to z_i.

We need to prove that this procedure terminates. If not, the exit test in Step 3 is never satisfied: $|z_i| < 1$ always, implying $\operatorname{Im} z_i < 1$ for all $i \geq 0$. By (7.9.1),

$$\operatorname{Im} z_{i+1} = \frac{\operatorname{Im} z_i}{|z_i|^2} > \operatorname{Im} z_i,$$

giving us a bounded increasing sequence $\operatorname{Im} z_0 < \operatorname{Im} z_1 < \cdots < 1$. For each i, $z_i = A_i z_0$ for some $A_i = \begin{bmatrix} j_i & k_i \\ l_i & m_i \end{bmatrix} \in \operatorname{SL}_2(\mathbb{Z})$. Then by (7.9.1) again,

$$\operatorname{Im} z_0 < \frac{\operatorname{Im} z_0}{|l_1 z_0 + m_1|^2} < \cdots < \frac{\operatorname{Im} z_0}{|l_i z_0 + m_i|^2} < \cdots < 1,$$

or, equivalently,

$$1 > |l_1 z_0 + m_1| > \cdots > |l_i z_0 + m_i| > \cdots \frac{1}{\sqrt{\operatorname{Im} z_0}}.$$

In geometric terms, we have found infinitely many points of the lattice $\mathbb{Z} z_0 + \mathbb{Z}$ in the annulus $(\operatorname{Im} z_0)^{-1/2} < z < 1$. This contradiction shows that the Reduction Algorithm for $\mathbb{C}$ indeed terminates, and that each point in $\mathbb{H}$ is $\operatorname{SL}_2(\mathbb{Z})$-equivalent to one in $\mathcal{F}_{\mathrm{im}}$.

To show that $\mathcal{F}_{\mathrm{im}}$ is the desired fundamental domain, it remains to show the following, for any $A = \begin{bmatrix} j & k \\ l & m \end{bmatrix} \in \operatorname{SL}_2(\mathbb{Z})$: if z and Az are both in $\mathcal{F}_{\mathrm{im}}$, then $z = Az$. Since $Az = (-A)z$, we may assume that either $l > 0$, or $l = 0, m > 0$. Switching z and Az if necessary, we may also assume that $\operatorname{Im} z \geq \operatorname{Im} Az$. Then (7.9.1) shows that $lz + m$ is in the closed half-disc $C = \{z \in \mathbb{H} : |z| \leq 1\}$. In other words, $lz \in l\mathcal{F}_{\mathrm{im}} \cap (C + \mathbb{Z})$, which is possible only in the following cases (draw a picture):

(a) $l = 0$. This forces $m = 1$, so $A = \begin{bmatrix} 1 & k \\ 0 & 1 \end{bmatrix}$ and $Az = z + k$. If z and $z + k$ were distinct elements of $\mathcal{F}_{\mathrm{im}}$, which has width 1, we'd have $k = \pm 1$. Then one of z and Az would lie on the line $\operatorname{Re} z = -1/2$, and the other on $\operatorname{Re} z = 1/2$. But the latter line is disjoint with $\mathcal{F}_{\mathrm{im}}$, hence $k = 0$ and $Az = z$.

(b) $l = 1$. Now $z \in \mathcal{F}_{\mathrm{im}}$ and $z + m \in C$, which happens only for $z = \zeta = e^{2\pi i/3}$ and $m = 1$. Then $A = \begin{bmatrix} j & j-1 \\ 1 & 1 \end{bmatrix}$ and $Az = (j\zeta + j - 1)/(\zeta + 1) = j + \zeta \in \mathcal{F}_{\mathrm{im}}$, which implies $j = 0$ and $Az = z$. ∎

7.9.3 Corollary. *The fundamental domain for the action of* $\operatorname{SL}_2(\mathbb{Z})$ *on* $\mathcal{H}_D$ *consists of quadratic numbers of discriminant* D *in* $\mathcal{F}_{\mathrm{im}}$.

Observe that the Reduction Algorithm for $\mathbb{H}$ parallels the Definite Reduction Algorithm of Prop. 7.1.6: if Step n of the latter transforms a form q into q', then Step n of the former takes its parameter η_q to $\eta_{q'}$.

7.9.4 Example. Take the sequence of forms of Ex. 7.1.8,

$$27x^2 - 95xy + 84y^2 \xrightarrow{T^{-2}} 27x^2 + 13xy + 2y^2$$

$$\xrightarrow{S} 2x^2 - 13xy + 27y^2 \xrightarrow{T^{-3}} 2x^2 - xy + 6y^2.$$

The corresponding parameters, depicted in Fig. 7.2,

$$\frac{95 + \sqrt{-47}}{54} \xrightarrow{T^{-2}} \frac{-13 + \sqrt{-47}}{54} \xrightarrow{S} \frac{13 + \sqrt{-47}}{4} \xrightarrow{T^{-3}} \frac{1 + \sqrt{-47}}{4},$$

form the sequence of numbers $z_0 \to z_1 \to z_2 \to z_3$ produced by the Reduction Algorithm for $\mathbb{H}$.   □

We formally recall the notion of a reduced positive definite form introduced in Ex. 7.1.9.

7.9.5 Definition. *Let* $q(x,y) = ax^2 + bxy + cy^2$ *be a positive definite quadratic form, so that* $b^2 - 4ac < 0, a > 0$. *The form* q *is said to be* **reduced** *if the following conditions hold:*

(a) $|b| \leq a \leq c$, *and*
(b) $b \geq 0$ *when* $a = |b|$ *or* $a = c$

7.9.6 Proposition. *A positive definite quadratic form* $q(x,y) = ax^2 + bxy + cy^2$ *is reduced if and only if* $\eta_q \in \mathcal{F}_{\mathrm{im}}$.

Proof. We have $\eta_q = (-b + i\sqrt{4ac - b^2})/2a \in \mathbb{H}$, so

$$\mathrm{Re}\,\eta_q = -\frac{b}{2a} \quad \text{and} \quad |\eta_q|^2 = \frac{b^2 + 4ac - b^2}{4a^2} = \frac{c}{a}.$$

The conditions $|\mathrm{Re}\,z| \leq 1/2$ and $|z| \geq 1$, which define the closure $\overline{\mathcal{F}_{\mathrm{im}}}$, translate to

$$\left| -\frac{b}{2a} \right| \leq \frac{1}{2}, \quad \frac{c}{a} \geq 1.$$

Multiplying by $a > 0$, we get the inequalities of Def. 7.9.5 (a). Condition (b) applies when η_q is on the boundary of $\overline{\mathcal{F}_{\mathrm{im}}}$: $a = |b|$ if and only if $\mathrm{Re}\,\eta_q = \pm 1/2$, and $a = c$ if and only if $|\eta_q| = 1$. Taking $b > 0$ corresponds, in both cases, to selecting the q with $\mathrm{Re}\,\eta_q < 0$.   ∎

The $\mathrm{SL}_2(\mathbb{Z})$-equivariant bijection $\iota_{\mathcal{H}\mathcal{Q}} : \mathcal{H}_D \to \mathcal{Q}_D$ translates the Proposition into the following result on forms, which we saw in Ex. 7.1.9 and Exer. 7.1.6 for discriminant -47.

7.9.7 Corollary. *A positive definite quadratic form is equivalent to a unique reduced form.*

To find the fundamental domain for the action of $\mathrm{SL}_2(\mathbb{Z})$ on $\mathcal{Q}_D$, we need a procedure for listing all reduced forms of discriminant D.

7.9.8 Proposition. *A complete list of reduced positive definite forms* $q(x,y) = ax^2 + bxy + cy^2$ *of discriminant* $D < 0$ *is determined by the following conditions on the coefficients* a, b, c:

(7.9.9) $$|b| \leq \sqrt{\frac{|D|}{3}}, \quad b \equiv D \pmod 2$$

(7.9.10) $$a_{\min} = |b| \leq a \leq \sqrt{\frac{b^2 - D}{4}} = a_{\max}$$

(7.9.11) $$c = \frac{b^2 - D}{4a} \in \mathbb{Z}$$

(7.9.12) $\quad b \geq 0 \ when \ a = |b| \ or \ a = c$

Proof. Assume that q is reduced. From $|b| \leq a \leq c$ we get $-D = 4ac - b^2 \geq 4 |b| \cdot |b| - |b|^2 = 3 |b|^2$. Moreover, $b \equiv b^2 \equiv D \pmod{2}$. The second inequality of (7.9.10) follows from $a^2 \leq ac = (b^2 - D)/4$. Finally, (7.9.11) limits us to those a for which c is also in $\mathbb{Z}$.

For the converse, take a form satisfying (7.9.9)–(7.9.12). All the conditions for being reduced are already on that list, except $a \leq c$. That we get by combining (7.9.11) with the second inequality of (7.9.10). ∎

Take $D = D_F$ for an imaginary quadratic field F. The proposition, combined with the identification $\mathrm{Cl}(F) \cong \mathcal{Q}_F / \mathrm{SL}_2(\mathbb{Z})$ of Thm. 7.8.1, lets us quickly compute the class number of F by counting the reduced forms of discriminant D_F.

7.9.13 Example. Let's compute $h(\mathbb{Q}[\sqrt{-231}])$ by enumerating all reduced quadratic forms of discriminant $D_{\mathbb{Q}[\sqrt{-231}]} = -231$. We first find the values of $|b|$ allowed by (7.9.9). Next, we find the prime factorization of the corresponding value of ac, and go through its factors a satisfying the bounds (7.9.10). We summarize the computation in the following table:

$|b| = a_{\min}$	$ac = \frac{b^2+231}{4}$	$\lfloor \sqrt{ac} \rfloor = a_{\max}$	(a, c)
1	$2 \cdot 29$	7	$(1, 58)^+, (2, 29)$
3	$2^2 \cdot 3 \cdot 5$	7	$(3, 20)^+, (4, 15), (5, 12), (6, 10)$
5	2^6	8	$(8, 8)^+$
7	$2 \cdot 5 \cdot 7$	8	$(7, 10)^+$

Each pair in the last column stands, in principle, for the two reduced forms $ax^2 \pm bxy + cy^2$. The exception are the pairs marked with a $+$. They contribute just the form with $b \geq 0$, either because they satisfy the boundary condition (7.9.12), or because $b = 0$ (which doesn't happen in this example). We see that the total number of forms is $4 + 2 \cdot 4 = 12$.

It's no wonder that this computation is easier than the ideal class group computations of Ch. 5: we merely determined the size of $\mathrm{Cl}(\mathbb{Q}[\sqrt{-231}])$, not its group structure. We still get some information about its 2-torsion. Any abelian group of order 12 is isomorphic to either $\mathbb{Z}/2\mathbb{Z} \times \mathbb{Z}/2\mathbb{Z} \times \mathbb{Z}/3\mathbb{Z}$, or $\mathbb{Z}/4\mathbb{Z} \times \mathbb{Z}/3\mathbb{Z}$. The two groups are distinguished by their 2-torsion, $\mathbb{Z}/2\mathbb{Z} \times \mathbb{Z}/2\mathbb{Z}$ or $\mathbb{Z}/2\mathbb{Z}$, respectively. By Exer. 7.9.4, ideal classes of order 2 correspond precisely to the forms marked with a $+$, so that

$$\mathrm{Cl}(\mathbb{Q}[\sqrt{-231}]) \cong \mathbb{Z}/2\mathbb{Z} \times \mathbb{Z}/2\mathbb{Z} \times \mathbb{Z}/3\mathbb{Z}. \qquad \square$$

Exercises

7.9.1. Apply the Reduction Algorithm for $\mathbb{H}$ to move the given point to the fundamental domain $\mathcal{F}_{\text{im}}$: (a) $(\pi + i)/10$; (b) $(-10 + \sqrt{-21})/9$; (c) $(-2 + 20i)/101$.

7.9.2. Show directly that $\mathcal{F}_{\text{im}}$ contains only finitely many quadratic numbers of any fixed negative discriminant.

7.9.3. Find all reduced forms of discriminant $D < 0$ that look like $x^2 + bxy + cy^2$. When $D = D_F$, which ideal classes in $\text{Cl}(F)$ do they correspond to?

7.9.4. Let F be an imaginary quadratic field. Let $(\mathscr{I}, \alpha, \beta)$ be an oriented ideal for which the corresponding form $q_{\mathscr{I}, \alpha, \beta}(x, y) = ax^2 + bxy + cy^2$ is reduced. Prove that $[\mathscr{I}]^2 = [\mathcal{O}]$ in $\text{Cl}(F)$ if and only if one of the following holds: $b = 0$, $a = |b|$, or $a = c$. In all three cases, $ax^2 + bxy + cy^2$ is the only reduced form with outer coefficients a and c.

7.9.5. Let $F = \mathbb{Q}[\sqrt{-D}]$ for $D \in \{5, 14, 23, 71, 89, 163\}$. Compute $h(F)$ by listing all reduced forms of discriminant D_F. Where possible, follow Ex. 7.9.13 to deduce restrictions on the structure of the 2-torsion of $\text{Cl}(F)$. Compare with your ideal class group computations in Ch. 5.

7.9.6. Find all matrices $A \in \text{SL}_2(\mathbb{Z})$ that satisfy $AX = XA$ for each $X \in \text{SL}_2(\mathbb{Z})$.

7.9.7. In the upper half-plane, draw the effect of the matrices $S = \left[\begin{smallmatrix} 0 & -1 \\ 1 & 0 \end{smallmatrix}\right]$ and $T = \left[\begin{smallmatrix} 1 & 1 \\ 0 & 1 \end{smallmatrix}\right]$ on $z \in \mathbb{H}$. For S, it's best to write z as $re^{i\varphi}$.

7.9.8. In this exercise, we prove that $\text{SL}_2(\mathbb{Z})$ is generated by the **standard generators** S and T:

$$\text{SL}_2(\mathbb{Z}) = \langle S, T \rangle = \{S^{a_1} T^{b_1} \cdots S^{a_n} T^{b_n} : a_i, b_i \in \mathbb{Z}\}.$$

(a) Let $z \in \mathbb{H}$. Show that the stabilizer subgroup $\text{SL}_2(\mathbb{Z})_z = \{A \in \text{SL}_2(\mathbb{Z}) : Az = z\}$ is equal to $\{\pm \left[\begin{smallmatrix} 1 & 0 \\ 0 & 1 \end{smallmatrix}\right]\}$, unless z is a quadratic number.

(b) Fix a $z \in \mathcal{F}_{\text{im}}$ which is not a quadratic number, and take any $A \in \text{SL}_2(\mathbb{Z})$. The Reduction Algorithm for $\mathbb{H}$ of Prop. 7.9.2 produces a $M \in \langle S, T \rangle$ for which $MAz \in \mathcal{F}_{\text{im}}$. Deduce from this that $A = \pm M \in \langle S, T \rangle$, so that S and T indeed generate $\text{SL}_2(\mathbb{Z})$.

7.9.9. We now prove the presentation

$$\text{SL}_2(\mathbb{Z}) = \langle S, T : S^2 = (ST)^3 = -\left[\begin{smallmatrix} 1 & 0 \\ 0 & 1 \end{smallmatrix}\right] \rangle$$

by showing that there are no additional relations between S and T.

(a) Verify the two relations.
(b) Put $U = ST$. Using $S^2 = U^3 = -\left[\begin{smallmatrix} 1 & 0 \\ 0 & 1 \end{smallmatrix}\right]$, show that any relation $S^{a_1} U^{b_1} \ldots S^{a_r} U^{b_r} = \left[\begin{smallmatrix} 1 & 0 \\ 0 & 1 \end{smallmatrix}\right]$ reduces to one of the form

$$(SU^{e_1})(SU^{e_2}) \cdots (SU^{e_t}) = \pm \left[\begin{smallmatrix} 1 & 0 \\ 0 & 1 \end{smallmatrix}\right], \text{ where } e_i = 1 \text{ or } 2.$$

(c) Both SU and SU^2 have one zero and three negative entries. Conclude that any product of at least two of them will have no more than one zero entry. The extra relation in (b) thus cannot hold.

7.9.10. Find a presentation (generators and relations) for $\mathrm{GL}_2(\mathbb{Z})$.

7.10 Reduced Indefinite Forms

The reduced quadratic forms of discriminant $D < 0$ form a fundamental domain for the action of $\mathrm{SL}_2(\mathbb{Z})$ on $\mathcal{Q}_D$. Geometrically, they are the quadratic forms with a parameter that lies in the fundamental domain $\mathcal{F}_{\mathrm{im}}$ of Fig. 7.2. We would like to have an analogous theory for indefinite forms.

The parameter of an indefinite quadratic form is a real quadratic number, and as such has a periodic continued fraction. The *purely* periodic quadratic numbers are, by Thm. 6.4.10, those $\eta \in \mathcal{H} \cap \mathbb{R}$ for which $\rho(\eta) = (\bar{\eta}, \eta)$ is in the region

$$\mathcal{F}_{\mathrm{re}} = \{(x,y) : -1 < x < 0, y > 1\} \subset \mathbb{R}^2.$$

This region is *not* a fundamental domain for some natural $\mathrm{SL}_2(\mathbb{Z})$-action on $\mathbb{R}^2$. The possibility that $\mathcal{F}_{\mathrm{re}}$ contains several points in the same $\mathrm{SL}_2(\mathbb{Z})$-orbit is an essential feature of the reduction theory for indefinite quadratic forms. Still, by formal analogy with Prop. 7.9.6, we make the following definition.

7.10.1 Definition. *An indefinite quadratic form q is* **reduced** *if its parameter is purely periodic, i.e., if $\eta_q \in \mathcal{F}_{\mathrm{re}}$.*

Here's a simple but useful consequence of the definition.

7.10.2 Proposition. *A reduced indefinite quadratic form $q(x,y) = ax^2 + bxy + cy^2$ has $a > 0$ and $b < 0$.*

Proof. Let $T \subset \mathbb{R}^2$ be the quadrant above the lines $y = -x$ and $y = x$. Since q is reduced, η_q is in $\mathcal{F}_{\mathrm{re}}$, which is visibly a subset of T (see Fig. 7.3). For any real quadratic η, Fig. 5.4 shows that $\rho(\eta) \in T$ if and only if $\mathrm{Ra}\,\eta, \mathrm{Ir}\,\eta > 0$. Thus

$$\mathrm{Ir}\,\eta_q = \sqrt{\frac{\mathrm{disc}\,q}{D_F}}\frac{1}{2a} > 0 \text{ and } \mathrm{Ra}\,\eta_q = \frac{-b}{2a} > 0.$$

We conclude that $a > 0$ and $b < 0$. ∎

Take an indefinite form with parameter η. The Continued Fraction Procedure of Sec. 6.1 defines the usual sequences $\{a_i\}$, $\{p_i\}$, $\{q_i\}$, and $\{\eta_i\}$, so that $\eta = [a_0, \ldots, a_{i-1}, \eta_i]$. Choose the smallest r and l such that $\eta_{i+l} = \eta_i$ for $i \geq r$.

7.10.3 Proposition. *The orbit $\mathrm{GL}_2(\mathbb{Z})\eta$ contains exactly l purely periodic continued fractions, namely the tails $\eta_r, \ldots, \eta_{r+l-1}$.*

Proof. Putting $M_i = \begin{bmatrix} q_{i-2} & -p_{i-2} \\ -q_{i-1} & p_{i-1} \end{bmatrix}$, we get from Lemma 6.3.1 that

$$(7.10.4) \qquad\qquad \eta_i = M_i \eta, \text{ with } \det M_i = (-1)^i.$$

Thus, all tails, periodic or not, are in the $\mathrm{GL}_2(\mathbb{Z})$-orbit of η.

Conversely, the next proposition shows that any $\eta' \in \mathrm{GL}_2(\mathbb{Z})\eta$ has the same tails as η: $\eta'_{j'_0+k} = \eta_{j_0+k}$ for some $j_0, j'_0 \geq 0$ and all $k \geq 0$. A purely periodic η' is itself a tail of η, since there is an m with $\eta' = \eta'_{j'_0+m} = \eta_{j_0+m}$. ∎

7.10.5 Proposition. *The continued fractions of $\eta, \eta' \in \mathbb{R} \setminus \mathbb{Q}$ have the same tails if and only if $\eta' \in \mathrm{GL}_2(\mathbb{Z})\eta$.*

Proof. We leave the "only if" part as an exercise, and just show that η and $A\eta$ have the same tails for $A \in \mathrm{GL}_2(\mathbb{Z})$. It's enough to check this when A ranges over a set of generators of $\mathrm{GL}_2(\mathbb{Z})$. One such set consists of $\begin{bmatrix} 1 & 1 \\ 0 & 1 \end{bmatrix}, \begin{bmatrix} -1 & 0 \\ 0 & 1 \end{bmatrix}$, and $\begin{bmatrix} 0 & 1 \\ 1 & 0 \end{bmatrix} = \begin{bmatrix} -1 & 0 \\ 0 & 1 \end{bmatrix}\begin{bmatrix} 0 & -1 \\ 1 & 0 \end{bmatrix}$: by Exer. 7.9.8, $\begin{bmatrix} 1 & 1 \\ 0 & 1 \end{bmatrix}$ and $\begin{bmatrix} 0 & -1 \\ 1 & 0 \end{bmatrix}$ generate $\mathrm{SL}_2(\mathbb{Z})$, while $\begin{bmatrix} -1 & 0 \\ 0 & 1 \end{bmatrix}$ is in the only nontrivial coset of $\mathrm{SL}_2(\mathbb{Z})$ in $\mathrm{GL}_2(\mathbb{Z})$. Applying any of these generators to η produces a continued fraction with the same tails as η:

$$\begin{bmatrix} 1 & 1 \\ 0 & 1 \end{bmatrix}\eta = \eta + 1 = [a_0 + 1, a_1, a_2, \dots]$$

$$\begin{bmatrix} -1 & 0 \\ 0 & 1 \end{bmatrix}\eta = -\eta = \begin{cases} [-(a_0+1), 1, a_1 - 1, a_2, a_3, \dots] & \text{if } a_1 \neq 1 \\ [-(a_0+1), a_2 + 1, a_3, a_4, \dots] & \text{if } a_1 = 1 \end{cases}$$

$$\begin{bmatrix} 0 & 1 \\ 1 & 0 \end{bmatrix}\eta = 1/\eta = [0, a_0, a_1, \dots] \text{ for } a_0 > 0$$

We can always ensure that $a_0 > 0$ by repeatedly applying $\begin{bmatrix} 1 & 1 \\ 0 & 1 \end{bmatrix}$. ∎

While $\mathrm{GL}_2(\mathbb{Z})$-equivalence is natural in the context of continued fractions, it is $\mathrm{SL}_2(\mathbb{Z})$-equivalence of quadratic forms and their parameters that we really care about, in light Thm. 7.8.1. By the determinant formula of (7.10.4), η_{2i} is $\mathrm{SL}_2(\mathbb{Z})$-equivalent to η. For i large enough, η_{2i} is purely periodic, and is therefore the parameter of a reduced form properly equivalent to q_η. Those forms and their parameters come in a periodic sequence with a natural ordering: $\eta_{2i}, \eta_{2i+2}, \eta_{2i+4}, \dots$. The η_{2i+1} may or may not be $\mathrm{SL}_2(\mathbb{Z})$-equivalent to η, depending on the parity of its period length.

7.10.6 Example. Let's find all reduced forms properly equivalent to $q(x,y) = x^2 - xy - 4y^2$, which has parameter $\eta = (1 + \sqrt{17})/2 = [2, \overline{1, 1, 3}]$. By Prop. 7.10.3, η_1, η_2, η_3 are all the distinct purely periodic continued fractions in the $\mathrm{GL}_2(\mathbb{Z})$-orbit of η (see Fig. 7.3). When $i \geq 1$, the η_{2i} are purely periodic and $\mathrm{SL}_2(\mathbb{Z})$-equivalent to η. But $\eta_4 = \eta_1, \eta_6 = \eta_3$, etc.: since the period length of η is odd, its $\mathrm{SL}_2(\mathbb{Z})$-orbit contains all its purely periodic tails. We list them in the order of appearance, along with the corresponding reduced forms:

$$\eta_2 = \overline{[1,3,1]} = \frac{1+\sqrt{17}}{4}, \qquad q_2(x,y) = 2x^2 - xy - 2y^2$$

$$\eta_4 = \overline{[1,1,3]} = \frac{3+\sqrt{17}}{4} = \eta_1, \qquad q_1(x,y) = 2x^2 - 3xy - y^2$$

$$\eta_6 = \overline{[3,1,1]} = \frac{3+\sqrt{17}}{2} = \eta_3, \qquad q_3(x,y) = x^2 - 3xy - 2y^2$$

Check that $h(\mathbb{Q}[\sqrt{17}]) = 1$ and $N\varepsilon_{\mathbb{Q}[\sqrt{17}]} = -1$. By Prop. 7.7.5 and Thm. 7.8.1, $\mathcal{Q}_{17}/\operatorname{SL}_2(\mathbb{Z}) \cong \operatorname{Cl}^+(\mathbb{Q}[\sqrt{17}]) = \operatorname{Cl}(\mathbb{Q}[\sqrt{17}]) = \{[\mathcal{O}]\}$, so any two forms of discriminant 17 are properly equivalent. The only reduced forms in the single equivalence class are q_2, q_1, and q_3. □

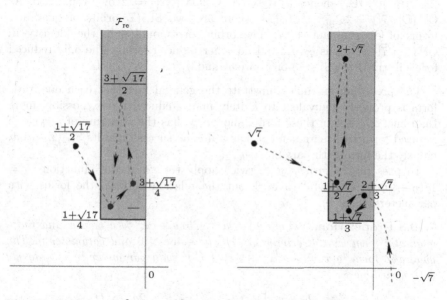

Fig. 7.3 Reduction of $x^2 - xy - 4y^2$ (*left*); $x^2 - 7y^2$ and $-x^2 + 7y^2$ (*right*). A *bold point* labeled α is $\rho(\alpha)$.

7.10.7 Example. Next, we list all reduced forms properly equivalent to the form $q(x,y) = x^2 - 7y^2$ of parameter $\eta = \sqrt{7} = [2, \overline{1,1,1,4}]$. For $i \geq 1$, η_i is purely periodic with period length 4, and the elements of $\mathcal{F}_{\mathrm{re}}$ that are $\operatorname{SL}_2(\mathbb{Z})$-equivalent to η are $\eta_2, \eta_4, \eta_6 = \eta_2, \eta_8 = \eta_4$, etc. We see that there are only two distinct purely periodic elements in the $\operatorname{SL}_2(\mathbb{Z})$-orbit of η, corresponding to two reduced forms:

$$\eta_2 = \overline{[1,1,4,1]} = \frac{1+\sqrt{7}}{2}, \quad q_2(x,y) = 2x^2 - 2xy - 3y^2$$

$$\eta_4 = \overline{[4,1,1,1]} = 2 + \sqrt{7}, \quad q_4(x,y) = x^2 - 4xy - 3y^2.$$

The remaining periodic tails η_1 and η_3 appear as the parameters of the reduced forms properly equivalent to $q'(x,y) = -q(x,y) = -x^2 + 7y^2$, the parameter of which is $\eta' = -\sqrt{7} = [-3, 2, \overline{1, 4, 1, 1}]$. Purely periodic numbers in the orbit $\mathrm{SL}_2(\mathbb{Z})\eta'$, and the corresponding forms, are

$$\eta_2' = [\overline{1,4,1,1}] = \eta_3 = \frac{1+\sqrt{7}}{3}, \quad q_2'(x,y) = 3x^2 - 2xy - 2y^2$$

$$\eta_4' = [\overline{1,1,1,4}] = \eta_1 = \frac{2+\sqrt{7}}{3}, \quad q_4'(x,y) = 3x^2 - 4xy - y^2$$

What is the connection with $\mathrm{Cl}^+(\mathbb{Q}[\sqrt{7}])$? Since the period length of $\sqrt{7}$ is even, Exer. 6.7.5 (a) shows that $N\varepsilon_{\mathbb{Q}[\sqrt{7}]} = 1$. You can check that $\mathbb{Z}[\sqrt{7}]$ is a PID, whence $h^+(\mathbb{Q}[\sqrt{7}]) = 2h(\mathbb{Q}[\sqrt{7}]) = 2$ by Prop. 7.7.5. As $\mathrm{Cl}^+(\mathbb{Q}[\sqrt{7}]) \cong \mathcal{Q}_{\mathbb{Q}[\sqrt{7}]}/\mathrm{SL}_2(\mathbb{Z})$, there are two $\mathrm{SL}_2(\mathbb{Z})$-orbits of quadratic forms of discriminant $4 \cdot 7$. The orbit corresponding to the identity in $\mathrm{Cl}^+(\mathbb{Q}[\sqrt{7}])$ contains q_2, q_4, and no other reduced forms. The only reduced forms in the other $\mathrm{SL}_2(\mathbb{Z})$-orbit are q_2' and q_4'. $\qquad\square$

The preceding examples illustrate the general fact that each quadratic form is properly equivalent to finitely many reduced forms, possibly more than one. The set of these forms, moreover, has the structure of a cycle: a reduced form with parameter η_i has a unique successor in its $\mathrm{SL}_2(\mathbb{Z})$-orbit, namely the form with parameter η_{i+2}.

To pass from η_i to η_{i+2} we twice apply the "next tail" function $^\sharp\eta = 1/(\eta - \lfloor\eta\rfloor)$. It's useful to work out the relation between the forms with parameters η and $^\sharp\eta$.

7.10.8 Proposition. *Let $q(x,y) = ax^2 + bxy + cy^2$ be a reduced indefinite quadratic form with discriminant $D = b^2 - 4ac > 0$ and parameter η. The quadratic form $^\sharp q(x,y) = a'x^2 + b'xy + c'y^2$ with parameter $^\sharp\eta$ is uniquely determined by*

$$b' \equiv -b \pmod{2a}, \quad -\sqrt{D} < b' < 2a - \sqrt{D}$$

$$c' = -a, \quad a' = \frac{b'^2 - D}{4a}.$$

Proof. First, let's check that these conditions determine a', b' and c' uniquely. There are precisely $2a$ integers in the interval $(-\sqrt{D}, 2a - \sqrt{D})$, among them only one $b' \equiv -b \pmod{2a}$. That congruence implies $b'^2 \equiv b^2 \pmod{4a}$, hence $b'^2 - D \equiv 0 \pmod{4a}$, and a' is indeed an integer.

To show that $^\sharp q$ has parameter $^\sharp\eta$, we verify the two conditions of Prop. 7.3.3. Put $s = \lfloor\eta\rfloor$. Since η is a root of $q(x,1)$, we know that $^\sharp\eta = 1/(\eta - s) = \left[\begin{smallmatrix} 0 & 1 \\ 1 & -s \end{smallmatrix}\right]\eta$ is a root of $\left(\left[\begin{smallmatrix} 0 & 1 \\ 1 & -s \end{smallmatrix}\right]q\right)(x,1)$. Therefore, up to sign, $^\sharp q(x,y)$ equals

$$\left(\left[\begin{smallmatrix} 0 & 1 \\ 1 & -s \end{smallmatrix}\right]q\right)(x,y) = q\left(\left[\begin{smallmatrix} s & 1 \\ 1 & 0 \end{smallmatrix}\right]\left[\begin{smallmatrix} x \\ y \end{smallmatrix}\right]\right) = q(sx + y, x)$$

(7.10.9)

$$= (as^2 + bs + c)x^2 + (b + 2as)xy + ay^2.$$

As η is purely periodic, we have $-1 < \bar{\eta} < 0 < s < \eta < s+1$. In particular, s lies between the x-intercepts $\bar{\eta}$ and η of the parabola $y = ax^2 + bx + c$. Since $a > 0$ by Prop. 7.10.2, we deduce that

$$\operatorname{Ir}{}^{\sharp}\eta = \frac{-1}{(\eta - s)(\bar{\eta} - s)} \operatorname{Ir}\eta > 0 \quad \text{and} \quad \pm a' = as^2 + bs + c = q(s,1) < 0.$$

To ensure $\operatorname{Ir}{}^{\sharp}\eta / a' > 0$, we must take the negative of the form (7.10.9):

$$^{\sharp}q(x,y) = -(as^2 + bs + c)x^2 + (-b - 2as)xy - ay^2.$$

From here we read off $c' = -a$ and $b' \equiv -b \pmod{2a}$. Solving the estimate

$$s < \eta = \frac{-b + \sqrt{D}}{2a} < s+1$$

for bounds on $b' = -b - 2as$ gives $-\sqrt{D} < b' < 2a - \sqrt{D}$. $\blacksquare$

The forms $q(x,y)$ and $^{\sharp}q(x,y)$ are related by the matrix $\begin{bmatrix} 0 & 1 \\ 1 & -s \end{bmatrix} \in \mathrm{GL}_2(\mathbb{Z}) \setminus \mathrm{SL}_2(\mathbb{Z})$, so they need not, on the face of it, be $\mathrm{SL}_2(\mathbb{Z})$-equivalent. By contrast, $q(x,y)$ and $^{\sharp\sharp}q(x,y)$ are always in the same $\mathrm{SL}_2(\mathbb{Z})$-orbit. This accounts for the musical notation $^{\sharp}q(x,y)$: to get from q to its successor in its $\mathrm{SL}_2(\mathbb{Z})$-orbit, we need to take two half-steps.

7.10.10 Example. Let's use Prop. 7.10.8 to re-discover the $\mathrm{SL}_2(\mathbb{Z})$-orbit of the reduced form $q_2(x,y) = 2x^2 - xy - 2y^2$ of discriminant 17. We have $^{\sharp}q_2(x,y) = a'x^2 + b'xy - 2y^2$, where

$$b' \equiv 1 \pmod 4 \quad \text{and} \quad \left\lceil -\sqrt{17} \right\rceil = -4 \le b' \le -1 = \left\lfloor 4 - \sqrt{17} \right\rfloor.$$

We must have $b' = -3$, so $^{\sharp}q_2(x,y) = x^2 - 3xy - 2y^2$. Iterating this procedure produces a sequence of forms, which we tabulate alternating between two columns. The arrows point from q to $^{\sharp}q$:

$q = {}^{\sharp}q'$	$q' = {}^{\sharp}q$
$q_2 = 2x^2 - xy - 2y^2$	$x^2 - 3xy - 2y^2 = q_3$
$q_1 = 2x^2 - 3xy - y^2$	$2x^2 - xy - 2y^2 = q_2$
$q_3 = x^2 - 3xy - 2y^2$	$2x^2 - 3xy - y^2 = q_1$

Passing from a form q to the one below it requires two half-steps, and yields the successor form $^{\sharp\sharp}q$. Each column therefore lists, in order, a complete $\mathrm{SL}_2(\mathbb{Z})$-orbit of forms. In our example, the two columns differ only by a shift, so there is a single $\mathrm{SL}_2(\mathbb{Z})$-orbit of reduced forms, the same one we found in Ex. 7.10.6. $\square$

To use reduction theory as a tool for computing class numbers of real quadratic fields, we need to systematically enumerate all indefinite reduced forms of a given discriminant.

7.10.11 Proposition. *An indefinite form* $q(x,y) = ax^2 + bxy + cy^2$ *of discriminant* $D = b^2 - 4ac > 0$ *is reduced if and only if*

$$\left|\sqrt{D} - 2a\right| < -b < \sqrt{D}.$$

Proof. Assume that q is reduced. Keeping in mind that $a > 0$, the condition $\eta_q = (-b + \sqrt{D})/2a \in \mathcal{F}_{\mathrm{re}}$ becomes

$$-2a < -b - \sqrt{D} < 0 < 2a < -b + \sqrt{D}.$$

The second inequality shows that $-b < \sqrt{D}$. Solving the first and the last inequality for bounds on $\sqrt{D}$, we get $b < 2a - \sqrt{D} < -b$, and finally $\left|\sqrt{D} - 2a\right| < -b$ as desired. For the converse, observe that $\left|\sqrt{D} - 2a\right| < \sqrt{D}$ implies $a > 0$, and reverse the argument. ∎

It is useful to re-write these bounds in a form closer to an algorithm.

7.10.12 Corollary. *The list of all reduced indefinite forms* $ax^2 + bxy + cy^2$ *of discriminant* $D = b^2 - 4ac > 0$ *is determined by the following conditions on the coefficients* a, b, c:

$$-\left\lfloor \sqrt{D} \right\rfloor \le b < 0, \ b \equiv D \pmod 2$$

$$a_{\min} = \left\lceil \frac{\sqrt{D} + b}{2} \right\rceil \le a \le \left\lfloor \frac{\sqrt{D} - b}{2} \right\rfloor = a_{\max}$$

$$c = \frac{b^2 - D}{4a} \in \mathbb{Z}.$$

7.10.13 Example. We will use the Corollary to produce the list of all reduced forms of discriminant $316 = 4 \cdot 79$. We summarize the computations in a table, dropping the values of b for which no a, c exist:

b	$ac = \frac{b^2 - 316}{4}$	$a_{\min}$	$a_{\max}$	(a, c)
-6	$-2 \cdot 5 \cdot 7$	6	12	$(7, -10), (10, -7)$
-8	$-3^2 \cdot 7$	5	13	$(7, -9), (9, -7)$
-10	$-2 \cdot 3^3$	4	14	$(6, -9), (9, -6)$
-14	$-2 \cdot 3 \cdot 5$	2	16	$(2, -15), (3, -10), (5, -6),$
				$(6, -5), (10, -3), (15, -2)$
-16	$-3 \cdot 5$	1	17	$(1, -15), (3, -5), (5, -3), (15, -1)$

We now sort these reduced forms into their $\mathrm{SL}_2(\mathbb{Z})$-orbits, in order of succession. We can start anywhere: for instance, $^\#(x^2 - 16xy - 15y^2) = a'x^2 + b'xy - y^2$ where $b' \equiv 16 \pmod 2$, $\left\lceil \sqrt{316} - 2 \right\rceil = 16 \le -b' \le \left\lfloor \sqrt{316} \right\rfloor = 17$, so that $b' = -16$ and $a' = 15$. Continuing, we tabulate the forms alternating between two columns, so that in each column q is followed by $^{\#\#}q$:

$$q = {}^\sharp q' \qquad\qquad\qquad q' = {}^\sharp q$$

$q = {}^\sharp q'$	$q' = {}^\sharp q$
$x^2 - 16xy - 15y^2$	$15x^2 - 16xy - y^2$
$2x^2 - 14xy - 15y^2$	$15x^2 - 14xy - 2y^2$

The calculation ${}^\sharp(15x^2 - 14xy - 2y^2) = x^2 - 16xy - 15y^2$ closes the cycle. We've found two $SL_2(\mathbb{Z})$-orbits of reduced forms, each with two elements.

Next, we take a form we haven't tabulated yet, like $7x^2 - 6xy - 10y^2$, and fill out a similar table of successors:

$q = {}^\sharp q'$	$q' = {}^\sharp q$
$7x^2 - 6xy - 10y^2$	$9x^2 - 8xy - 7y^2$
$6x^2 - 10xy - 9y^2$	$5x^2 - 14xy - 6y^2$
$3x^2 - 16xy - 5y^2$	$10x^2 - 14xy - 3y^2$

Again, we get two $SL_2(\mathbb{Z})$-orbits, each of length 3. The analogous table starting with $10x^2 - 6xy - 7y^2$ finally exhausts the list of reduced forms of discriminant 316:

$q = {}^\sharp q'$	$q' = {}^\sharp q$
$10x^2 - 6xy - 7y^2$	$3x^2 - 14xy - 10y^2$
$5x^2 - 16xy - 3y^2$	$6x^2 - 14xy - 5y^2$
$9x^2 - 10xy - 6y^2$	$7x^2 - 8xy - 9y^2$

Check that $\mathrm{Cl}(\mathbb{Q}[\sqrt{79}]) = \mathbb{Z}/3\mathbb{Z}$ and $\mathrm{Cl}^+(\mathbb{Q}[\sqrt{79}]) = \mathbb{Z}/6\mathbb{Z}$. The six narrow ideal classes correspond to the six $SL_2(\mathbb{Z})$-orbits of reduced forms we computed. $\qquad\square$

Exercises

7.10.1.* Show that an indefinite quadratic form $q(x,y) = ax^2 + bxy + cy^2$ is reduced if and only if $\left|\sqrt{D} + 2c\right| < -b < \sqrt{D}$.

7.10.2. Show directly that $x^2 - 7y^2$ and $-x^2 + 7y^2$ are not $GL_2(\mathbb{Z})$-equivalent. How is this fact related to the properties of the fundamental unit of $\mathbb{Q}[\sqrt{7}]$?

7.10.3. The forms q_1 and q_3 of Ex. 7.10.6 are obviously related by $\left[\begin{smallmatrix}0&1\\1&0\end{smallmatrix}\right] \in GL_2(\mathbb{Z}) \setminus SL_2(\mathbb{Z})$. Find an $A \in SL_2(\mathbb{Z})$ for which $q_3 = Aq_1$.

7.10.4. Find all reduced forms $SL_2(\mathbb{Z})$-equivalent to: (a) $3x^2 - 16xy - 6y^2$; (b) $x^2 + 5xy + 3y^2$; (c) $-2x^2 + 7xy + y^2$; (d) $5x^2 - 5xy + 2y^2$.

7.10.5. For $D \in \{19, 85, 223, 235, 401\}$, list all reduced forms of discriminant D. Arrange them in cycles, as in Ex. 7.10.13, and use this to determine $h^+(\mathbb{Q}[\sqrt{D}])$.

7.10.6. Let $q(x,y) = ax^2 + bxy + cy^2$ and ${}^\sharp q(x,y) = a'x^2 + b'xy + c'y^2$. Show that ${}^\sharp q = \left[\begin{smallmatrix}0&1\\1&\frac{b+b'}{2a}\end{smallmatrix}\right] q$.

7.11 Form Composition and Bhargava Cubes

Let F be an arbitrary quadratic field, and $(\mathscr{I}, \alpha, \beta)$ an oriented fractional ideal. The $\mathrm{SL}_2(\mathbb{Z})$-equivalence class $[q_{\mathscr{I},\alpha,\beta}]$ doesn't depend on the choice of oriented basis (α, β); we abbreviate it to $[q_{\mathscr{I}}]$. By Thm. 7.8.1, the map $[\mathscr{I}] \mapsto [q_{\mathscr{I}}]$ is a bijection $\iota_{\mathcal{IQ}} : \mathrm{Cl}^+(F) \cong \mathcal{Q}_F / \mathrm{SL}_2(\mathbb{Z})$. There is a unique group structure on $\mathcal{Q}_F / \mathrm{SL}_2(\mathbb{Z})$ which makes $\iota_{\mathcal{IQ}}$ into a group isomorphism. It is given by the following operation.

7.11.1 Definition. *The **composition** of two $\mathrm{SL}_2(\mathbb{Z})$-equivalence classes of quadratic forms in $\mathcal{Q}_F / \mathrm{SL}_2(\mathbb{Z})$ is given by*

$$[q_{\mathscr{I}_1}][q_{\mathscr{I}_2}] = [q_{\mathscr{I}_1 \cdot \mathscr{I}_2}].$$

We refer to $\mathcal{Q}_F / \mathrm{SL}_2(\mathbb{Z})$ under the composition as the **form class group of** F. Its identity is $[q_{\mathcal{O}}] = [x^2 - txy + ny^2]$, usually denoted simply by 1. Here we put $\mathcal{O} = \mathbb{Z}[\delta]$ with $\delta^2 - t\delta + n = 0$. By Exer. 7.8.6, the inverse of a class is given by

$$[ax^2 + bxy + cy^2]^{-1} = [ax^2 - bxy + cy^2] = [cx^2 + bxy + ay^2].$$

In this section, whenever we write an ideal as $\mathbb{Z}\alpha + \mathbb{Z}\beta$, we implicitly assume that the basis (α, β) is oriented.

7.11.2 Example. The ring of integers of $F = \mathbb{Q}[\sqrt{-23}]$ is $\mathcal{O} = \mathbb{Z}[\delta]$ with $\delta^2 - \delta + 6 = 0$. Consider its ideals $I = \mathbb{Z}\cdot 2 + \mathbb{Z}(-1+\delta)$ and $I^2 = \mathbb{Z}\cdot 4 + \mathbb{Z}(1+\delta)$. The square of the quadratic form $q(x,y) = q_{I,2,-1+\delta}(x,y) = 2x^2 + xy + 3y^2$ is, by definition,

$$[q]^2 = [q_{I^2,4,5+\delta}] = [4x^2 - 3xy + 2y^2] = [2x^2 - xy + 3y^2].$$

The last form is the reduced representative of its equivalence class. As $[q]^{-1} = [2x^2 - xy + 3y^2] = [q]^2$, we deduce that $[q]^3 = 1$. This agrees with Exer. 5.4.6 (e), where we found that $\mathrm{Cl}(\mathbb{Q}[\sqrt{-23}]) \cong \mathbb{Z}/3\mathbb{Z}$.

Putting $X = -xz + xw + yz + 2yw$ and $Y = -xz - xw - yz + yw$, a computer algebra program will check the following identity:

$$(2x^2 + xy + 3y^2)(2z^2 + zw + 3w^2) = 2X^2 - XY + 3Y^2.$$

In English, the product of two numbers "of the form" $2x^2 + xy + 3y^2$ is "of the form" $2x^2 - xy + 3y^2$. It is in this elementary guise that Legendre first discovered the composition of forms; it is also the origin of the term "form." The identity is unsurprising if we think of quadratic forms as associated to ideals, since it boils down to

$$\frac{N\alpha_1}{NI} \cdot \frac{N\alpha_2}{NI} = \frac{N(\alpha_1\alpha_2)}{N(I^2)}, \text{ for all } \alpha_1, \alpha_2 \in I. \qquad \square$$

We want a general recipe for composing the classes of two quadratic forms of discriminant D_F. The following lemma allows us to assume that they have relatively prime x^2-coefficients.

7.11.3 Lemma. *For $d \in \mathbb{N}$ and a quadratic form q, there exists a quadratic form $a'x^2 + b'xy + c'y^2$ properly equivalent to q, for which $\gcd(a', d) = 1$.*

Proof. It suffices to construct relatively prime integers l and m for which $\gcd(q(l, m), d) = 1$. Once we have them, we can find $r, s \in \mathbb{Z}$ with $rm - sl = 1$. Then the form $\left(\left[\begin{smallmatrix} r & -s \\ -l & m \end{smallmatrix} \right] q \right)(x, y) = q(l, m)x^2 + \cdots$ is properly equivalent to q, and its x^2-coefficient is relatively prime to d.

We look for l and m among the divisors of d. We group the prime factors p_i of d into three pairwise relatively prime numbers:

$$k = \prod_{p_i | \gcd(a,c)} p_i, \quad l = \prod_{p_i \nmid c} p_i, \quad m = \prod_{p_i \nmid a, \, p_i | c} p_i.$$

We will show that no p_i divides $q(l, m) = al^2 + blm + cm^2$. It's enough to check that two terms in the sum are divisible by p_i, while the third isn't. Each p_i divides precisely one of k, l and m, so we consider three cases:

$p_i \mid k$: As p_i divides a and c, it can't divide b, because the form q is primitive. Thus p_i divides the outer terms in $q(l, m)$, but not blm.

$p_i \mid l$: As $p_i \mid al^2 + blm$, it suffices to show that $p_i \nmid cm^2$. By definition, $p_i \nmid c$, and l is relatively prime to m.

$p_i \mid m$: Similarly, $p_i \nmid al^2$, while $p_i \mid blm + cm^2$. ∎

Take two forms of discriminant D_F, say $q_i(x, y) = a_ix^2 + b_ixy + c_iy^2$, for $i = 1, 2$. By the Lemma, we may assume that $\gcd(a_1, a_2) = 1$. By passing to an $SL_2(\mathbb{Z})$-equivalent form, e.g., a reduced one, we may assume that $a_1, a_2 > 0$. It's easy to check that the set $I_i = \mathbb{Z}a_i + (-b_i + \sqrt{D_F})/2$ is an oriented ideal for which the associated form is q_i. Since $NI_1 = a_1$ is relatively prime to $NI_2 = a_2$, the ideals I_1 and I_2 are themselves relatively prime, as in Ex. 4.6.4. In that case, an explicit formula for $I_1 I_2$ allows us to compute $[q_1][q_2]$.

7.11.4 Proposition. *There exist $b \in \mathbb{Z}$ satisfying $b \equiv b_i \pmod{2a_i}$ for $i = 1, 2$. For any such b, we have*

$$I_1 I_2 = \mathbb{Z} \cdot a_1 a_2 + \mathbb{Z} \frac{-b + \sqrt{D_F}}{2}.$$

Proof. Since $b_1 \equiv D_F \equiv b_2 \pmod 2$ and $\gcd(a_1, a_2) = 1$, the strong version of the Chinese Remainder Theorem (Exer. 1.1.11) produces $b \in \mathbb{Z}$ satisfying $b \equiv b_i \pmod{2a_i}$ for $i = 1, 2$. We have $2a_i \mid b - b_i$ and $2 \mid b + b_i$, so $4a_i \mid b^2 - b_i^2$, and

$$a_i \mid \frac{b^2 - b_i^2}{4} + \frac{b_i^2 - D_F}{4} = \frac{b^2 - D_F}{4} = N\left(\frac{-b + \sqrt{D_F}}{2} \right).$$

Since a_1 and a_2 are relatively prime, $a_1 a_2$ also divides $N((-b_i + \sqrt{D_F})/2)$, making $I = \mathbb{Z} \cdot a_1 a_2 + \mathbb{Z}(-b + \sqrt{D_F})/2$ an ideal.

Put $b = b_i - 2a_i k_i$ for some $k_i \in \mathbb{Z}$. Since $(-b + \sqrt{D_F})/2 = a_i k_i + (-b_i + \sqrt{D_F})/2 \in I_i$, we have that $I \subseteq I_1 \cap I_2 = I_1 I_2$, the last equality holding because I_1 and I_2 are relatively prime. Comparing norms, we deduce that $I = I_1 I_2$. ∎

The quadratic form corresponding to $I_1 I_2$, with the indicated oriented basis, is $(a_1 a_2)x^2 + bxy + cy^2$ for some $c \in \mathbb{Z}$. A linear change of variables in q_i produces a form whose xy-coefficient is also b:

$$(T^{k_i} q_i)(x, y) = a_i x^2 + (b_i - 2a_i k_i)xy + (a_i k_i^2 - b_i k_i + c_i)y^2$$
$$= a_i x^2 + bxy + c_i' y^2.$$

The three forms $a_1 x^2 + bxy + c_1' y^2$, $a_2 x^2 + bxy + c_2' y^2$, and $(a_1 a_2)x^2 + bxy + cy^2$ all have discriminant $D_F = b^2 - 4a_1 c_2' = b^2 - 4a_1 c_2' = b^2 - 4a_1 a_2 c$. We deduce that $c_1' = ca_2$ and $c_2' = ca_1$.

We assumed that $a_1, a_2 > 0$ for convenience only. You may check that the argument proves the following proposition regardless of signs.

7.11.5 Proposition. *Any two* $\mathrm{SL}_2(\mathbb{Z})$-*equivalence classes of quadratic forms of discriminant* D_F *have representatives* $a_1 x^2 + bxy + c_1 y^2$ *and* $a_2 x^2 + bxy + c_2 y^2$, *with* $\gcd(a_1, a_2) = 1$. *We say that two such quadratic forms are* **united**. *Put* $c = \gcd(c_1, c_2)$. *Then* $c_1 = ca_2$, $c_2 = ca_1$, *and the composition of the classes is given by*

$$[a_1 x^2 + bxy + (ca_2)y^2][a_2 x^2 + bxy + (ca_1)y^2] = [(a_1 a_2)x^2 + bxy + cy^2].$$

7.11.6 Example. Let's compose the classes of the forms

$$q_1(x, y) = 2x^2 - xy + 6y^2 \quad \text{and} \quad q_2(x, y) = 3x^2 + xy + 4y^2,$$

both of discriminant -47. To apply the proposition, we find that $b = 7$ satisfies the congruences $b \equiv -1 \pmod 4, b \equiv 1 \pmod 6$. Then

$$(T^{-2} q_1)(x, y) = 2x^2 + 7xy + 12y^2, \quad (T^{-1} q_2)(x, y) = 3x^2 + 7xy + 8y^2$$

are two united forms representing $[q_1]$ and $[q_2]$, respectively. Their composition is

$$[2x^2 - xy + 6y^2][3x^2 + xy + 4y^2]$$
$$= [2x^2 + 7xy + 12y^2][3x^2 + 7xy + 8y^2]$$
$$= [6x^2 + 7xy + 4y^2] = [3x^2 - xy + 4y^2].$$

The last form is the reduced representative of its class. Check this computation against the table in Fig. 7.1. □

Early in the twenty-first century, Bhargava discovered a new perspective on the 200-year-old theory of composition of forms. To motivate it, observe that F is a two-dimensional vector space over $\mathbb{Q}$, and fix a basis $\mathcal{B} = \{\beta_1, \beta_2\}$. Multiplication by $\alpha \in F$ is a $\mathbb{Q}$-linear transformation of F, whose matrix $M_{\mathcal{B}}(\alpha) = \begin{bmatrix} j & k \\ l & m \end{bmatrix}$ relative to $\mathcal{B}$ is determined by $\alpha\beta_1 = j\beta_1 + l\beta_2$ and $\alpha\beta_2 = k\beta_1 + m\beta_2$. The function $\alpha \mapsto M_{\mathcal{B}}(\alpha)$ preserves addition and multiplication. In Exer. 4.1.1, we saw that $\mathrm{N}\alpha = \det M_{\mathcal{B}}(\alpha)$ and $\mathrm{Tr}\,\alpha = \mathrm{tr}\,M_{\mathcal{B}}(\alpha)$.

7.11.7 Example. Let $I = \mathbb{Z}a + \mathbb{Z}(-b+\delta)$ be an ideal, so that $b^2 - tb + n = ja$ for some $j \in \mathbb{Z}$. Putting $\mathcal{B} = \{a, -b+\delta\}$, we easily compute

$$M_{\mathcal{B}}(a) = \begin{bmatrix} a & 0 \\ 0 & a \end{bmatrix}, \quad M_{\mathcal{B}}(-b+\delta) = \begin{bmatrix} 0 & -j \\ a & -2b+t \end{bmatrix}.$$

We express the form $q_{I,a,-b+\delta}$ in matrix terms:

$$q_{I,a,-b+\delta}(x,y) = \frac{\mathrm{N}(xa - y(-b+\delta))}{a} = \frac{\det\left(x \begin{bmatrix} a & 0 \\ 0 & a \end{bmatrix} - y \begin{bmatrix} 0 & -j \\ a & -2b+t \end{bmatrix}\right)}{a}$$

$$= \det\left(x \begin{bmatrix} 1 & 0 \\ 0 & a \end{bmatrix} - y \begin{bmatrix} 0 & -j \\ 1 & -2b+t \end{bmatrix}\right).$$

The last equality holds since both matrices on its left side have first column divisible by a.

We can recover $q_{I,a,-b+\delta}$ from the pair of matrices $\begin{bmatrix} 1 & 0 \\ 0 & a \end{bmatrix}, \begin{bmatrix} 0 & -j \\ 1 & -2b+t \end{bmatrix}$. Changing the typography slightly, those two matrices can be thought of as the front and back faces of a cube:

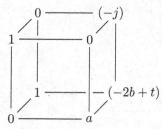

Such cubes are the protagonists of Bhargava's take on the composition of forms.

7.11.8 Definition. *A* **Bhargava cube** $\mathcal{C}_{abcdefgh}$ *is a* $2 \times 2 \times 2$ *array of elements in* $\mathbb{Z}$:

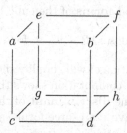

There are three ways of slicing a Bhargava cube into a pair of matrices. Given a subscript $* \in \{F, L, T\}$, let M_* be the matrix on the *Front*, *Left*, or *Top* face of the cube with a as its upper-left entry. Let N_* be the parallel matrix on the opposite face. The following table gives the three slicings. For orientation, we draw an arrow along the top row of M_* and N_*:

$*$	Slicing	$M_*,\quad N_*$

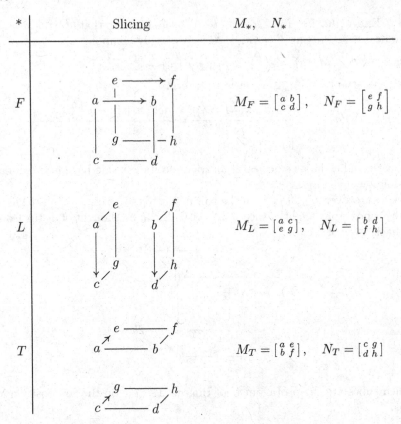

F $M_F = \begin{bmatrix} a & b \\ c & d \end{bmatrix}, \quad N_F = \begin{bmatrix} e & f \\ g & h \end{bmatrix}$

L $M_L = \begin{bmatrix} a & c \\ e & g \end{bmatrix}, \quad N_L = \begin{bmatrix} b & d \\ f & h \end{bmatrix}$

T $M_T = \begin{bmatrix} a & e \\ b & f \end{bmatrix}, \quad N_T = \begin{bmatrix} c & g \\ d & h \end{bmatrix}$

We refer to the cube by any of the symbols $\lVert M_F|N_F\rVert = \lVert M_L|N_L\rVert = \lVert M_T|N_T\rVert$. We always write the position subscripts on the matrices.

We define the **associated forms** of the cube by

$$q_*(x,y) = -\det(xM_* - yN_*), \text{ for } * \in \{F,L,T\}.$$

The minus sign in the definition of will simplify the connection between cubes and composition. A tedious computation shows that q_F, q_L, and q_T have the same discriminant, called the **discriminant** of the cube.

A cube is called **primitive** if all three forms q_F, q_L and q_T are primitive.

7.11.9 Example. Consider the cube

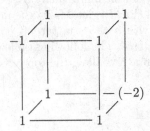

It's symmetric with respect to rotation by $2\pi/3$ around the diagonal connecting -1 and -2. That means that the three associated forms are equal: $q_F(x,y) = q_L(x,y) = q_T(x,y) = 2x^2 + xy + 3y^2$. This is the form q of Ex. 7.11.2. There we established that $[q]^3 = 1$, which we now re-write as $[q_F][q_L][q_T] = 1$.   □

7.11.10 Example. Take $a_1, a_2, b, c \in \mathbb{Z}$, with $\gcd(a_1, a_2) = 1$ and $b^2 - 4a_1a_2c = D_F$. Consider the following cube and its three associated forms:

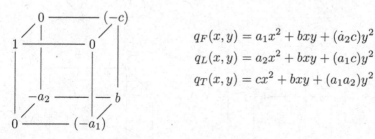

$$q_F(x,y) = a_1x^2 + bxy + (a_2c)y^2$$
$$q_L(x,y) = a_2x^2 + bxy + (a_1c)y^2$$
$$q_T(x,y) = cx^2 + bxy + (a_1a_2)y^2$$

This cube is primitive since $\gcd(a_1a_2, b, c) = 1$, which follows from $\gcd(a_1, a_2) = 1$ and the (near-) square-freenes of D_F.

The formula for composing united forms of Prop. 7.11.5 translates into $[q_F][q_L][q_T] = 1$. Indeed, the forms q_F and q_L are the two forms on the left side of the formula, while its right side is the inverse of $[q_T]$: $[cx^2 + bxy + (a_1a_2)y^2]^{-1} = [(a_1a_2)x^2 + bxy + cy^2]$.   □

The pattern in these examples persists: any Bhargava cube of discriminant D_F imposes a relation $[q_F][q_L][q_T] = 1$ in the group $\mathcal{Q}_F / \mathrm{SL}_2(\mathbb{Z})$. We will prove this by transforming an arbitrary cube into one from Ex. 7.11.10, without changing the classes $[q_F], [q_L]$ or $[q_T]$. For this we'll use three-dimensional analogues of row and column operations.

7.11.11 Definition. *Let* $* \in \{F, L, T\}$, *and let* $\mathrm{SL}_2(\mathbb{Z})_*$ *be a copy of* $\mathrm{SL}_2(\mathbb{Z})$. *Take* $A \in \mathrm{SL}_2(\mathbb{Z})$, *and denote by* A_* *the corresponding matrix in* $\mathrm{SL}_2(\mathbb{Z})_*$. *We define the (left) action of* A_* *on a primitive cube* $C = \lVert M_F | N_F \rVert = \lVert M_L | N_L \rVert = \lVert M_T | N_T \rVert$ *by*

$$A_* C = \lVert AM_* | AN_* \rVert.$$

The three actions of $\mathrm{SL}_2(\mathbb{Z})$ on the cube C with vertices $a, b, \ldots, h$ are determined by the actions of the generators $\begin{bmatrix} 1 & 1 \\ 0 & 1 \end{bmatrix}$ and $\begin{bmatrix} 0 & -1 \\ 1 & 0 \end{bmatrix}$ (see Exer. 7.9.8), as given in the following table:

The first cube in the table shows that any one of the following computes the action of $\begin{bmatrix} 1 & 1 \\ 0 & 1 \end{bmatrix}_F$:

$$A = \begin{bmatrix} 1 & 1 \\ 0 & 1 \end{bmatrix} \qquad\qquad A = \begin{bmatrix} 0 & -1 \\ 1 & 0 \end{bmatrix}$$

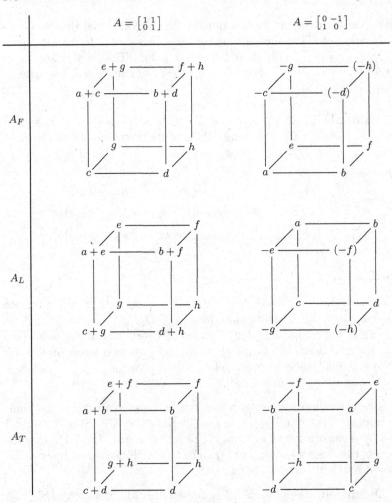

$$\begin{bmatrix} 1 & 1 \\ 0 & 1 \end{bmatrix}_F \mathcal{C} = \big[\!\big[\begin{bmatrix} 1 & 1 \\ 0 & 1 \end{bmatrix} M_F \,\big|\, \begin{bmatrix} 1 & 1 \\ 0 & 1 \end{bmatrix} N_F \big]\!\big] \quad \text{(add row 2 to row 1, in both } M_F \text{ and } N_F\text{)}$$

$$= \big[\!\big[M_L \begin{bmatrix} 1 & 0 \\ 1 & 1 \end{bmatrix} \,\big|\, N_L \begin{bmatrix} 1 & 0 \\ 1 & 1 \end{bmatrix} \big]\!\big] \quad \text{(add col. 2 to col. 1, in both } M_L \text{ and } N_L\text{)}$$

$$= \big[\!\big[M_T + N_T \,\big|\, N_T \big]\!\big].$$

Similar identities hold for any matrix in $\mathrm{SL}_2(\mathbb{Z})_*$. In particular, for $A \in \mathrm{SL}_2(\mathbb{Z})$ and its transpose A^t, we have

$$A_F \mathcal{C} = \big[\!\big[M_L A^t \,\big|\, N_L A^t \big]\!\big], \quad A_L \mathcal{C} = \big[\!\big[M_T A^t \,\big|\, N_T A^t \big]\!\big], \quad A_T \mathcal{C} = \big[\!\big[M_F A^t \,\big|\, N_F A^t \big]\!\big].$$

7.11.12 Proposition. *Any two of the actions of* $\mathrm{SL}_2(\mathbb{Z})_F$, $\mathrm{SL}_2(\mathbb{Z})_L$, *and* $\mathrm{SL}_2(\mathbb{Z})_T$ *commute, and therefore define an action of* $\mathrm{SL}_2(\mathbb{Z})_F \times \mathrm{SL}_2(\mathbb{Z})_L \times \mathrm{SL}_2(\mathbb{Z})_T$ *on primitive cubes.*

Proof. By definition, $\mathrm{SL}_2(\mathbb{Z})_L$ acts on the cube $\mathcal{C}$ by left-multiplying M_L and N_L, which performs row operations on both of them. Similarly, $\mathrm{SL}_2(\mathbb{Z})_F$ acts by column operations on M_L and N_L. Since row and column operations commute, so do the two actions. Formally, for $A, B \in \mathrm{SL}_2(\mathbb{Z})$, we have

$$A_L(B_F\mathcal{C}) = A_L \lVert M_L B^t | N_L B^t \rVert = \lVert AM_L B^t | AN_L B^t \rVert$$
$$= B_F \lVert AM_L | AN_L \rVert = B_F(A_L\mathcal{C}).$$

The commutativity of the other two pairs of actions is analogous. By Exer. 7.5.11, the three actions then combine into an action of $\mathrm{SL}_2(\mathbb{Z})_F \times \mathrm{SL}_2(\mathbb{Z})_L \times \mathrm{SL}_2(\mathbb{Z})_T$, given by $(A_F, B_L, C_T)\mathcal{C} = A_F(B_L(C_T\mathcal{C}))$. Since the actions commute, A_F, B_L and C_T can act on $\mathcal{C}$ in any order. ∎

Let's take $A \in \mathrm{SL}_2(\mathbb{Z})$ and determine how the action of A_F on a primitive cube $\mathcal{C}$ changes its associated forms q_F, q_L, and q_T. Marking with a prime the quantities associated to $A_F\mathcal{C}$, we find:

$$q_F'(x,y) = -\det(xM_F' - yN_F') = -\det(xAM_F - yAN_F)$$
$$= -\det A \cdot \det(xM_F - yN_F) = q_F(x,y)$$

$$q_L'(x,y) = -\det(xM_L' - yN_L') = -\det(xM_L A^t - yN_L A^t)$$
$$= -\det(xM_L - yN_L) \cdot \det A^t = q_L(x,y).$$

To see what happens to q_T, we compute with the standard generators of $\mathrm{SL}_2(\mathbb{Z})$. Since $\left[\begin{smallmatrix} 1 & 1 \\ 0 & 1 \end{smallmatrix}\right]_F \lVert M_T | N_T \rVert = \lVert M_T + N_T | N_T \rVert$, we find that

$$q_T'(x,y) = -\det(x(M_T + N_T) - yN_T) = -\det(xM_T - (y-x)N_T)$$
$$= q_T(x, y-x) = (\left[\begin{smallmatrix} 1 & 0 \\ 1 & 1 \end{smallmatrix}\right] q_T)(x,y).$$

Similarly, $\left[\begin{smallmatrix} 0 & -1 \\ 1 & 0 \end{smallmatrix}\right]_F \lVert M_T | N_T \rVert = \lVert -N_T | M_T \rVert$, and

$$q_T' = -\det(-xN_T - yM_T) = q_T(-y, x) = (\left[\begin{smallmatrix} 0 & 1 \\ -1 & 0 \end{smallmatrix}\right] q_T)(x,y).$$

We see that any $A_F \in \mathrm{SL}_2(\mathbb{Z})_F$ transforms q_T into an $\mathrm{SL}_2(\mathbb{Z})$-equivalent form. A similar calculation for the other two actions gives the following proposition.

7.11.13 Proposition. *Let* $M \in \mathrm{SL}_2(\mathbb{Z})_F \times \mathrm{SL}_2(\mathbb{Z})_L \times \mathrm{SL}_2(\mathbb{Z})_T$, *let* $\mathcal{C}$ *be a primitive cube, and take* $* \in \{F, L, T\}$. *Denote by* q_*' *the quadratic form associated to* $M\mathcal{C}$. *Then* $[q_*] = [q_*']$.

We're ready to prove the main theorem on Bhargava cubes.

7.11.14 Theorem. *The three forms associated with any primitive cube* $\mathcal{C}$ *of discriminant* D_F *satisfy* $[q_F][q_L][q_T] = 1$.

Proof. We will use the action of $SL_2(\mathbb{Z})_F \times SL_2(\mathbb{Z})_L \times SL_2(\mathbb{Z})_T$ to simplify $\mathcal{C}$ without changing the classes $[q_F]$, $[q_L]$, or $[q_T]$.

By a slight variation on Thm. 3.4.3, we can find matrices $X, Y \in GL_2(\mathbb{Z})$ for which $X M_F Y = \left[\begin{smallmatrix} g & 0 \\ 0 & s \end{smallmatrix}\right]$ with $g|s$, so that g is the g.c.d. of the entries of M_F. Since $\mathcal{C}$ is primitive and all the coefficients of q_L and q_T are divisible by g, we must have $g = 1$. After possibly multiplying by $\left[\begin{smallmatrix} 1 & 0 \\ 0 & -1 \end{smallmatrix}\right]$, thus changing the sign of s, we may assume that $X, Y \in SL_2(\mathbb{Z})$. The cube $(X_F, Y_L^t, \mathrm{id})\|M_F|N_F\|$ has the form of the first cube in the following diagram:

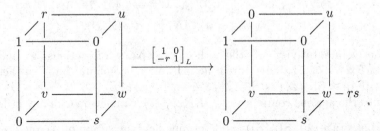

Subtracting r times the front of the first cube from its back produces the second cube. It has the same form as the cube in Ex. 7.11.10, with $a_1 = -s, a_2 = -v, c = -u$, and $h = w - rs$. The form $q_L'(x, y) = -vx^2 + (w - rs)xy + (su)y^2$ associated to the second cube is still primitive, hence $\gcd(a_1, a_2) = \gcd(s, v) = 1$. The conclusion of Ex. 7.11.10, and the invariance of the classes $[q_*]$ under the actions of the $SL_2(\mathbb{Z})_*$, imply that

$$1 = [a_1 x^2 + bxy + (a_2 c)y^2][a_2 x^2 + bxy + (a_1 c)y^2][cx^2 + bxy + (a_1 a_2)y^2]$$
$$= [q_F][q_L][q_T]. \quad \blacksquare$$

Theorem 7.11.14 allows us to compose the classes of any two forms if we can construct a suitable Bhargava cube.

7.11.15 Proposition. *Let q_1 and q_2 be two quadratic forms of the same discriminant. There exists a Bhargava cube for which $q_F = q_1$ and $q_L = q_2$.*

Proof. Put $q_i(x, y) = a_i x^2 + b_i xy + c_i y^2$ for $i = 1, 2$. To reduce the number of variables, we look for a cube with 0 at its front-left-bottom vertex:

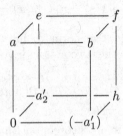

We want to choose the values at the vertices so that

$$q_F(x,y) = (aa_1')x^2 + (ah - a_1'e + a_2'b)xy - (eh + a_2'f)y^2 = q_1(x,y)$$
$$q_L(x,y) = (aa_2')x^2 + (ah + a_1'e - a_2'b)xy - (bh + a_1'f)y^2 = q_2(x,y).$$

Equating the x^2-coefficients and averaging the xy-coefficients forces

(7.11.16) $\qquad a_1 = aa_1', \quad a_2 = aa_2', \quad \dfrac{b_1 + b_2}{2} = ah.$

This suggests that we put $a = \gcd(a_1, a_2, (b_1 + b_2)/2)$, and define a_1', a_2' and h accordingly.

Equating the y^2-coefficients shows that f satisfies the system of congruences

(7.11.17) $\qquad a_2'f \equiv -c_1 \pmod{h}, \quad a_1'f \equiv -c_2 \pmod{h}.$

As disc $q_1 = $ disc q_2, we have the chain of implications

$$b_1^2 - 4a_1c_1 = b_2^2 - 4a_2c_2 \;\Rightarrow\; a_1c_1 \equiv a_2c_2 \pmod{(b_1 + b_2)/2} \;\Rightarrow$$
$$a_1'c_1 \equiv a_2'c_2 \pmod{h} \;\Rightarrow\; \gcd(a_2', h) \mid a_1'c_1.$$

By construction, $\gcd(a_1', a_2', h) = 1$, hence $\gcd(a_2', h) \mid c_1$, and similarly $\gcd(a_1', h) \mid c_2$. Put $d_i = \gcd(a_i', h), a_i'' = a_i'/d_i, h_i = h/d_i, c_1' = c_1/d_2$, and $c_2' = c_2/d_1$. Dividing each of the congruences (7.11.17) by the appropriate d_i, and observing that $\gcd(a_i'', h_i) = 1$, we see that the system (7.11.17) is equivalent to

$$f \equiv -a_2''^{-1}c_1' \pmod{h_2}, \quad f \equiv -a_1''^{-1}c_2' \pmod{h_1}.$$

The Chinese Remainder Theorem produces a solution f, as $\gcd(h_1, h_2) = \gcd(a_1', a_2', h) = 1$. Any such f will yield a cube we want. If $h \neq 0$, comparing the y^2-coefficients gives $e = -(c_1 + a_2'f)/h$ and $b = -(c_2 + a_1'f)/h$.

Except for the last sentence, the above argument works even when $h = 0$. In that case, all the congruences become equalities, and $f = -c_1/a_2' = -c_2/a_1'$. Comparing the y^2-coefficients yields no information on b and e. Instead, we chose them to be any solution to $-a_1'e + a_2'b = b_1$, which exists since $\gcd(a_1', a_2') = \gcd(a_1', a_2', 0) = 1$. ∎

7.11.18 Example. Let's follow the proof of Prop. 7.11.15 to compose the classes of the forms $q_1(x,y) = 2x^2 + 2xy + 17y^2$ and $q_2(x,y) = 6x^2 + 6xy + 7y^2$, both of discriminant -132. As in (7.11.16), we find that

$$a = \gcd(2, 6, (2 + 6)/2) = 2, \text{ so } a_1' = 1, a_2' = 3, \text{ and } h = 2.$$

By inspection, $f = 1$ satisfies the congruences (7.11.17),

$$3f \equiv -17 \pmod{2}, \quad f \equiv -7 \pmod{2}.$$

This choice of f is not unique, and neither is the cube we're constructing; this is why we only define the composition of classes of forms, and not of forms themselves. Finally, we compute $e = -(17 + 3 \cdot 1)/2 = -10$, and $b = -(7+1\cdot1)/2 = -4$. We have our cube, and the three associated quadratic forms:

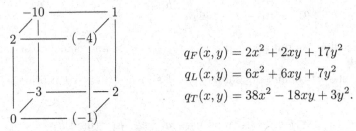

$$q_F(x,y) = 2x^2 + 2xy + 17y^2$$
$$q_L(x,y) = 6x^2 + 6xy + 7y^2$$
$$q_T(x,y) = 38x^2 - 18xy + 3y^2.$$

By Thm. 7.11.14, we find that

$$[q_1][q_2] = [q_F][q_L] = [q_T]^{-1} = [3x^2 - 18xy + 38y^2] = [3x^2 + 11y^2].$$

In fact, $[2x^2 + 2xy + 17y^2]$, $[6x^2 + 6xy + 7y^2]$, and $[3x^2 + 11y^2]$ are the three nonzero classes in $\mathcal{Q}_{\mathbb{Q}[\sqrt{-33}]}/\operatorname{SL}_2(\mathbb{Z}) \cong \operatorname{Cl}(\mathbb{Q}[\sqrt{-33}]) \cong \mathbb{Z}/2\mathbb{Z} \times \mathbb{Z}/2\mathbb{Z}$.   □

7.11.19 Example. Let's find the formula for squaring the class of a quadratic form by constructing the Bhargava cube when $q_1 = q_2$. Here $a_1 = a_2$; in a sense, this example is the opposite of Ex. 7.11.10, where $\gcd(a_1, a_2) = 1$.

Assume first that $b_1 \neq 0$. The equations (7.11.16) become

$$a = \gcd(a_1, b_1), \quad a_1' = a_2' = a_1/a, \quad h = b_1/a.$$

Fix $k, l \in \mathbb{Z}$ such that $ka_1' + lh = \gcd(a_1', h) = 1$. In particular, $a_1'^{-1} \equiv k$ (mod h). The congruences (7.11.17) reduce to $f \equiv -a_1'^{-1}c_1$ (mod h), and we may take $f = -c_1 k$. We now have to put $b = e = -(c_1 - a_1'c_1 k)/h = -c_1(1 - ka_1')/h = -c_1 lh/h = -c_1 l$. The desired cube is

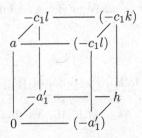

We invert q_T by switching the outside coefficients, and get that

$$[q_1]^2 = [q_T]^{-1} = [a_1'^2 x^2 + (b_1 - 2a_1'c_1 l)xy + (c_1 ak + c_1^2 l^2)y^2].$$

When $b_1 = 0$, check that $[a_1 x^2 + c_1 y^2]^2 = 1$.   □

In this section we only considered quadratic forms of discriminant D_F. All our results, however, hold for quadratic forms of arbitrary discriminant $D = c^2 D_F$, once we define their composition. For this we use, as in Def. 7.11.1, the natural bijection $\mathcal{Q}_D / \mathrm{SL}_2(\mathbb{Z}) \cong \mathrm{Cl}^+(\mathcal{O}_c)$. The latter group, defined just before Exer. 5.1.10, is the narrow class group of the order $\mathcal{O}_c = \mathbb{Z}[c\delta]$. Alternatively, the statements of Prop. 7.11.5 and Thm. 7.11.14 make sense for forms and cubes of arbitrary discriminant, and we could take either of them as the definition of composition. If we do that, we have to check that the composition is well-defined, and that it gives $\mathcal{Q}_D / \mathrm{SL}_2(\mathbb{Z})$ a group structure.

Exercises

7.11.1. Find the following compositions of classes in $\mathcal{Q}_F / \mathrm{SL}_2(\mathbb{Z})$, using both united forms and Bhargava cubes:

(a) $[7x^2 - 6xy - 10y^2][5x^2 - 14xy - 6y^2]$
(b) $[15x^2 - 10xy - 7y^2][7x^2 - 18xy - 7y^2]$
(c) $[6x^2 + 2xy + 9y^2][3x^2 + 2xy + 18y^2]$

7.11.2. Practice form composition in its elementary guise.

(a) Prove that

$$(2x^2 + 2xy + 3y^2)(2z^2 + 2zw + 3w^2) = X^2 + 5Y^2,$$

for some functions X and Y f the form $axz + bxw + cyz + dyw$.
(b) Express the product $(3x^2 + 2xy + 10y^2)(5z^2 + 2zw + 6w^2)$ in a similar way.

7.11.3. Let $q(x,y) = ax^2 + bxy + cy^2$ be a quadratic form of discriminant D_F. Show that q is associated to the oriented ideal $\mathbb{Z}a + \mathbb{Z}(-b + \sqrt{D_F})/2$ when $a > 0$, and $\sqrt{D_F}(\mathbb{Z}a + \mathbb{Z}(-b + \sqrt{D_F})/2)$ when $a < 0$. Use this to show that Prop. 7.11.5 is true as stated, regardless of the signs of a_1, a_2.

7.11.4. Prove that the three forms associated with the cube $\mathcal{C}_{abcdefgh}$ have the same discriminant, given by

$$\mathrm{disc}(q_F) = \mathrm{disc}(q_L) = \mathrm{disc}(q_T) = a^2 h^2 + b^2 g^2 + c^2 f^2 + d^2 e^2$$
$$- 2(abgh + cdef + acfh + bdeg + aedh + bfcg)$$
$$+ 4(adfg + bceh).$$

7.11.5. Construct Bhargava cubes that show that:

(a) $[ax^2 + bxy + cy^2][ax^2 - bxy + cy^2] = 1$;
(b) $[ax^2 + bxy + cy^2][cx^2 + bxy + ay^2] = 1$

7.11.6. Let q_F, q_L and q_T be the forms associated to a Bhargava cube.

(a) Show that if two of the forms q_F, q_L and q_T are primitive, then so is the third.

(b) Give an example of a cube where q_F is primitive, but q_L and q_T are not.
(c) Referring to the forms of Ex. 7.11.10, show that if $\gcd(a_1, a_2) = 1$ and q_T is primitive, then so are q_F and q_L.

7.11.7. Check that the identity class, in the appropriate $\mathcal{Q}_F/\operatorname{SL}_2(\mathbb{Z})$, is the square of each of the classes $[ax^2+cy^2]$, $[ax^2+axy+cy^2]$, and $[ax^2+bxy+ay^2]$. Compare with the exceptional cases of the Definite Reduction Algorithm of Prop. 7.1.6, and with the forms marked with a "+" in Ex. 7.9.13.

7.11.8. Generalize Ex. 7.11.9 to construct infinitely many quadratic fields F for which $3 \mid h(F)$.

7.11.9. Let $\mathcal{B}$ be the set of all Bhargava cubes, and let $R : \mathcal{B} \to \mathcal{B}$ be the rotation by $2\pi/3$ about the ah diagonal. Denote by $q_*^{\mathcal{C}}$ the forms associated to a cube $\mathcal{C}$. Show that $q_L^{\mathcal{C}} = q_F^{R\mathcal{C}}$ and $q_T^{\mathcal{C}} = q_F^{R^2\mathcal{C}}$.

7.11.10. Let $\mathcal{G}_D$ be the free abelian group generated by quadratic forms of discriminant D:

$$\mathcal{G}_D = \left\{ \sum_{i=1}^n a_i q_i : a_i \in \mathbb{Z}, q_i \in \mathcal{Q}_D \right\}.$$

Here we treat the q_i as symbols, so that all sums are formal. Consider the subgroup $\mathcal{R}_D \subseteq \mathcal{G}_D$ generated by all linear combinations of the form $q_F + q_L + q_T$, where the q_* are the three quadratic forms associated to some Bhargava cube of discriminant D. Put $(q) = q + \mathcal{R}_D \in \mathcal{G}_D/\mathcal{R}_D$.

(a) The action of $\operatorname{SL}_2(\mathbb{Z})_F \times \operatorname{SL}_2(\mathbb{Z})_L \times \operatorname{SL}_2(\mathbb{Z})_T$ on $\mathcal{Q}_D$ extends by linearity to $\mathcal{G}_D$. Show that this action induces a trivial action on $\mathcal{G}_D/\mathcal{R}_D$: $(Mq) = (q)$ for all $M \in \operatorname{SL}_2(\mathbb{Z})_F \times \operatorname{SL}_2(\mathbb{Z})_L \times \operatorname{SL}_2(\mathbb{Z})_T$.
(b) Show that $(q) \mapsto [q]$ is a well-defined group isomorphism $\mathcal{G}_D/\mathcal{R}_D \xrightarrow{\sim} \mathcal{Q}_D/\operatorname{SL}_2(\mathbb{Z})$.

Orders

7.11.11. Let $D = c^2 D_F$ be an arbitrary discriminant. Define the composition of classes of forms of discriminant D via the isomorphism $\mathcal{Q}_D/\operatorname{SL}_2(\mathbb{Z}) \cong \operatorname{Cl}^+(\mathcal{O}_c)$. Check that the results of this chapter remain valid in this more general context.

Appendix

Let F be any quadratic field, real or complex, and $\mathcal{O} = \mathbb{Z}[\delta]$ its ring of integers, with $\delta^2 - t\delta + n = 0$. In this appendix we collect the results on the arithmetic of subrings of $\mathcal{O}$, developed in the exercises.

A.1 Proposition (Exer. 4.2.11). *Any subring of $\mathcal{O}$, apart from $\mathbb{Z}$, is of the form $\mathcal{O}_c = \mathbb{Z} + \mathbb{Z}c\delta = \mathbb{Z}[c\delta]$ for a unique $c \in \mathbb{N}$. We call $\mathcal{O}_c$ the* **order of conductor** *c. In fact, $\mathcal{O}_c = \mathbb{Z}[\delta_c]$ for any $\delta_c \in \mathcal{O}$ satisfying an equation $\delta_c^2 - t_c\delta_c + n_c = 0$ with $t_c^2 - 4n_c = c^2 D_F = \operatorname{disc} \mathcal{O}_c$. We have that $\mathcal{O}_c \subseteq \mathcal{O}_{c'}$ if and only if $c' \mid c$.*

A.2 Example. We will illustrate the claims in this appendix with the order of discriminant $43,708 = 7^2 \cdot 2^2 \cdot 223 = c^2 D_F$. Given the constraints on the discriminant of a field, this is only possible when $c = 7$, $D_F = 4 \cdot 223$. Thus, $F = \mathbb{Q}[\sqrt{223}]$, our old friend from Exs. 5.4.12 and 6.6.5, and the order is $\mathcal{O}_7 = \mathbb{Z}[7\sqrt{223}]$. $\qquad\square$

We start by describing the units in $\mathcal{O}_c$.

A.3 Proposition (Exer. 6.7.6). *Assume that $c > 1$, i.e., $\mathcal{O}_c \subsetneq \mathcal{O}$. If F is imaginary, then $\mathcal{O}_c^\times = \{\pm 1\}$.*

If F is real with fundamental unit ε_F, then $\mathcal{O}_c^\times = \pm\langle \varepsilon_F^s \rangle = \{\pm \varepsilon_F^{sk} : k \in \mathbb{Z}\}$. The exponent s has the following equivalent characterizations:

(a) $s = [\mathcal{O}^\times : \mathcal{O}_c^\times]$
(b) s is the smallest positive exponent for which $\varepsilon_F^s \in \mathcal{O}_c$.
(c) s is the smallest positive integer for which $c \mid q_{sl-1}$. Here l is the period length of the continued fraction of δ, and p_i/q_i its ith convergent.

A.4 Example. Let's find the fundamental unit $\varepsilon_{\mathcal{O}_7}$. In Ex. 6.7.5 we found that $\varepsilon_F = 224 + 15\sqrt{223}$ is the fundamental unit of $\mathcal{O}$. Then $\varepsilon_F^2 = 100,351 + 6,720\sqrt{223} \in \mathcal{O}_7$, as $6,720 = 7 \cdot 960$. We conclude that $s = 2$ and $\mathcal{O}_c^\times = \pm\langle \varepsilon_F^2 \rangle$. $\qquad\square$

In Prop. 4.5.1 we constructed the standard form of an ideal of $\mathcal{O}$. The argument works when we replace δ by $c\delta$, giving the analogous standard form for ideals in orders.

M. Trifković, *Algebraic Theory of Quadratic Numbers*, Universitext,
DOI 10.1007/978-1-4614-7717-4, © Springer Science+Business Media New York 2013

A.5 Proposition (Exer. 4.5.4). *A subset $I \subseteq \mathcal{O}_c$ is an ideal of $\mathcal{O}_c$ if and only if we can find $a, b, d \in \mathbb{Z}$ satisfying*

$$I = d(\mathbb{Z}a + \mathbb{Z}(-b + c\delta)) \text{ with } b^2 - (ct)b + c^2 n \equiv 0 \pmod{a}.$$

Such an expression is called a **standard form of I relative to $\mathcal{O}_c$**. The latter qualifier is important, as the following example illustrates.

A.6 Example (Exer. 4.5.5). The standard form of the ideal $c\mathcal{O}$ relative to $\mathcal{O}_c$ is $\mathbb{Z} \cdot c + \mathbb{Z} \cdot c\delta$, with $a = c$, $b = 0$ and $d = 1$. Viewing $c\mathcal{O}$ as an ideal of the bigger ring $\mathcal{O}$, the standard form becomes $c\mathcal{O} = c(\mathbb{Z} + \mathbb{Z}\delta)$, with $a = 1$, $b = 0$, and $d = c$.

As this example shows, an ideal of I of $\mathcal{O}_c$ may also be an ideal of a bigger order. The ideals for which that doesn't happen have particularly nice arithmetic.

A.7 Definition. *Let I be an ideal of $\mathcal{O}_c$. The **norm of I relative to $\mathcal{O}_c$** is* $\mathrm{N}_c I = |\mathcal{O}_c / I|$.

A.8 Proposition (Exer. 4.6.8). *Let $I = d(\mathbb{Z}a + \mathbb{Z}(-b + c\delta))$ be an ideal of $\mathcal{O}_c$ in standard form. The following hold:*

(a) $\mathrm{N}_c I = d^2 a$.
(b) $\mathrm{N}_c(\mathcal{O}_c \alpha) = \mathrm{N}(\mathcal{O}\alpha) = |\mathrm{N}\alpha|$ for any $\alpha \in \mathcal{O}_c$.
(c) If $\gcd(a, c) = 1$, we have $I \cdot \bar{I} = \mathcal{O}_c \cdot \mathrm{N}_c I$, as in Thm. 4.6.5.

Thanks to claim (c), the proof of unique factorization for ideals of $\mathcal{O}$ carries over to ideals of $\mathcal{O}_c$ which have norm relatively prime to c.

A.9 Proposition. *Let I be an ideal of $\mathcal{O}_c$ of norm prime to c. Then $I = P_1 \cdots P_r$, where P_i are prime ideals of $\mathcal{O}_c$ of norm prime to c. The P_i are unique up to ordering.*

A.10 Proposition (Exer. 4.6.10). *Let $I = d(\mathbb{Z}a + \mathbb{Z}(-b + c\delta))$ be an ideal of $\mathcal{O}_c$. The following statements are equivalent:*

(a) $\gcd(a, c) = 1$.
(b) I can always be cancelled: if J, K are ideals of $\mathcal{O}_c$, then $IJ = IK$ implies $J = K$.
(c) I doesn't absorb multiplication by any order bigger than $\mathcal{O}_c$.

An ideal satisfying these conditions is called an **invertible ideal of $\mathcal{O}_c$**. Condition (c) has a useful rephrasing in terms of lattices. Here, we slightly abuse the term "lattice" to refer to any subgroup of F of the form $\Lambda = \mathbb{Z}\alpha + \mathbb{Z}\beta$ with $\beta/\alpha \notin \mathbb{Q}$. There is no risk of confusion: embedding F into a plane turns Λ into a lattice in the sense of Def. 3.1.1. Orders appear naturally as symmetry rings of lattices.

A.11 Definition. *The **ring of multipliers** of Λ is $\mathcal{O}_\Lambda = \{\gamma \in F : \gamma\Lambda \subseteq \Lambda\}$.*

A.12 Proposition (Exer. 4.2.13). *Let $\Lambda \subset F$ be a lattice. Let $j, k, l, m, c \in \mathbb{Z}$ be the unique quintuple for which $c > 0$, $\gcd(j, k, l, m, c) = 1$, and*

$$\delta\alpha = \tfrac{j}{c}\alpha + \tfrac{k}{c}\beta, \quad \delta\beta = \tfrac{l}{c}\alpha + \tfrac{m}{c}\beta.$$

Then $\mathcal{O}_\Lambda = \mathcal{O}_c$.

Condition (c) of Prop. A.10 inspires the following definition.

A.13 Definition. *A **fractional ideal** for $\mathcal{O}_c$ is a lattice $\mathscr{I} \subset F$ for which $\mathcal{O}_{\mathscr{I}} = \mathcal{O}_c$.*

The set $\mathbb{I}_{\mathcal{O}_c}$ of all fractional ideals of $\mathcal{O}_c$ is a group under multiplication. Inverses exist by virtue of Prop. A.8 (c). The subgroup of principal fractional ideals is $\mathbb{P}_{\mathcal{O}_c} = \{\mathcal{O}_c\alpha : \alpha \in F^\times\}$. Their quotient is a natural generalization of $\mathrm{Cl}(F)$.

A.14 Definition. *The **ideal class group** of $\mathcal{O}_c$ is $\mathrm{Cl}(\mathcal{O}_c) = \mathbb{I}_{\mathcal{O}_c}/\mathbb{P}_{\mathcal{O}_c}$. The **class number** of $\mathcal{O}_c$ is $h(\mathcal{O}_c) = |\mathrm{Cl}(\mathcal{O}_c)|$.*

We will soon see that $\mathrm{Cl}(\mathcal{O}_c)$ is finite. The same quotient can be obtained in two easier ways: first by replacing $I_{\mathcal{O}_c}$ and $\mathbb{P}_{\mathcal{O}_c}$ with more manageable subgroups, and then by identifying those with subgroups of $\mathbb{I}_F$. To do this we need some technical preliminaries. For $e = 1$ or c, put

$$\mathbb{I}_e(c) = \{I_1 I_2^{-1} : I_1, I_2 \text{ ideals of } \mathcal{O}_e, \gcd(N_e I_1, c) = \gcd(N_e I_2, c) = 1\},$$
$$F_e^\times(c) = \{\alpha/\beta : \alpha, \beta \in \mathcal{O}_e, \gcd(N\alpha, c) = \gcd(N\beta, c) = 1\}$$
$$\mathbb{P}_e(c) = \{\mathcal{O}_e\gamma : \gamma \in F_c^\times(c)\}$$

Some remarks:

(a) We have the redundant notations $\mathbb{I}_F = \mathbb{I}_{\mathcal{O}_1} = \mathbb{I}_1(1)$ and $\mathbb{P}_F = \mathbb{P}_{\mathcal{O}_1} = \mathbb{P}_1(1)$.
(b) $\mathbb{I}_c(c)$ is the smallest subgroup of $\mathbb{I}_{\mathcal{O}_c}$ containing all (necessarily invertible) ideals of $\mathcal{O}_c$ of norm prime to c. It is generated by the prime ideals of $\mathcal{O}_c$ not dividing c. Similar statements hold for $\mathbb{I}_1(c)$, viewed as a subgroup of $\mathbb{I}_F$.
(c) $\mathbb{P}_1(c)$ (resp. $\mathbb{P}_c(c)$) is the smallest subgroup of $\mathbb{I}_1(c)$ (resp. $\mathbb{I}_c(c)$) containing the principal ideals with generators that are in $\mathcal{O}_c$, *in both cases*. This equality of the sets of generators is key to the second statement of the following proposition.

A.15 Proposition (Exers. 4.7.9 and 4.6.13). *The assignment $\mathscr{I} \mapsto \mathscr{I}\mathcal{O}$ defines a group isomorphism $\mathbb{I}_c(c) \cong \mathbb{I}_1(c)$. This isomorphism identifies $\mathbb{P}_c(c)$ with $\mathbb{P}_1(c)$.*

The inverse isomorphism sends an (integral) ideal J of $\mathcal{O}$ to the ideal $J \cap \mathcal{O}_c$ of $\mathcal{O}_c$.

A.16 Proposition. *We have group isomorphisms*

$$\mathrm{Cl}(\mathcal{O}_c) = \mathbb{I}_{\mathcal{O}_c}/\mathbb{P}_{\mathcal{O}_c} \cong \mathbb{I}_c(c)/\mathbb{P}_c(c) \cong \mathbb{I}_1(c)/\mathbb{P}_1(c).$$

Proof. The second isomorphism follows directly from Prop. A.15. For the first isomorphism, observe that by the equivalence of Prop. A.10 (a) and (c), each invertible ideal of $\mathcal{O}_c$ can be scaled to one of norm prime to c. In other words, $\mathbb{I}_{\mathcal{O}_c} = \mathbb{I}_c(c)\mathbb{P}_{\mathcal{O}_c}$. The Second Isomorphism Theorem for groups gives us the identifications

$$\mathrm{Cl}(\mathcal{O}_c) = \mathbb{I}_{\mathcal{O}_c}/\mathbb{P}_{\mathcal{O}_c} = (\mathbb{I}_c(c)\mathbb{P}_{\mathcal{O}_c})/\mathbb{P}_{\mathcal{O}_c} \cong \mathbb{I}_c(c)/(\mathbb{I}_c(c) \cap \mathbb{P}_{\mathcal{O}_c}) = \mathbb{I}_c(c)/\mathbb{P}_c(c).\ \blacksquare$$

Take $\alpha \in \mathcal{O}$, and denote by $\tilde{\alpha}$ its coset in $\mathcal{O}/c\mathcal{O}$. It's easy to check that $\tilde{\alpha} \in (\mathcal{O}/c\mathcal{O})^\times$ if and only if $\gcd(\mathrm{N}\alpha, c) = 1$. We put $\widetilde{\alpha/\beta} = \tilde{\alpha}\tilde{\beta}^{-1}$ for all $\alpha, \beta \in \mathcal{O}$ with $\gcd(\mathrm{N}\alpha, c) = \gcd(\mathrm{N}\beta, c) = 1$. This extends reduction modulo c to a multiplicative homomorphism $F_1^\times(c) \to (\mathcal{O}/c\mathcal{O})^\times$. Further composing with the quotient homomorphism $(\mathcal{O}/c\mathcal{O})^\times \to (\mathcal{O}/c\mathcal{O})^\times/(\mathbb{Z}/c\mathbb{Z})^\times$, we get a group homomorphism $\varphi : F_1^\times(c) \to (\mathcal{O}/c\mathcal{O})^\times/(\mathbb{Z}/c\mathbb{Z})^\times$ with kernel $F_c^\times(c)$.

A.17 Proposition (Exers. 5.1.10–5.1.13). *The assignment* $\mathbb{P}_1(c)\cdot\mathscr{I} \mapsto \mathbb{P}_F\cdot\mathscr{I}$ *defines a surjective homomorphism*

$$\mathrm{Cl}(\mathcal{O}_c) \cong \mathbb{I}_1(c)/\mathbb{P}_1(c) \to \mathbb{I}_F/\mathbb{P}_F = \mathrm{Cl}(F)$$

with kernel isomorphic to $\mathbb{P}_F/\mathbb{P}_1(c) \cong ((\mathcal{O}/c\mathcal{O})^\times/(\mathbb{Z}/c\mathbb{Z})^\times)/\varphi(\mathcal{O}^\times)$.

For any field F we have that $\mathcal{O}^\times = \pm\langle\varepsilon_F\rangle$ (if F is imaginary, we can take $\varepsilon_F = \pm 1$, with the exceptions $\varepsilon_{\mathbb{Q}[i]} = i, \varepsilon_{\mathbb{Q}[\sqrt{-3}]} = (1 + \sqrt{-3})/2$). Then $|\varphi(\mathcal{O}^\times)|$ is the order of $\varphi(\varepsilon_F)$ in $(\mathcal{O}/c\mathcal{O})^\times/(\mathbb{Z}/c\mathbb{Z})^\times$. When F is real, that order is $[\mathcal{O}^\times : \mathcal{O}_c^\times]$ (see Prop. A.3).

A.18 Example. As $\left(\frac{223}{7}\right) = -1$, the prime 7 is inert in $F = \mathbb{Q}[\sqrt{223}]$, and Exer. 4.9.12 shows that $(\mathcal{O}/7\mathcal{O})^\times/(\mathbb{Z}/7\mathbb{Z})^\times$ is cyclic of order 8. The preceding discussion and Prop. A.3 give $|\varphi(\mathcal{O}^\times)| = 2$, so that the kernel of $\mathrm{Cl}(\mathcal{O}_7) \to \mathrm{Cl}(F)$ is cyclic of order $8/2 = 4$. The class number of $\mathcal{O}_7$ is thus $h(\mathcal{O}_7) = 4h(\mathcal{O}) = 12$. $\square$

Hints to Selected Exercises

1.1.8. Observe that $f(x) = x^2$ is a homomorphism of multiplicative groups $\mathbb{Z}/p\mathbb{Z} \setminus 0 \to \mathbb{Z}/p\mathbb{Z} \setminus 0$, find its kernel, then count.

1.1.9. From the first congruence we get that $x = a_1 + tn_1$ for some $t \in \mathbb{Z}$. Plug this into the second congruence and solve for t, which is possible because $\gcd(n_1, n_2) = 1$.

1.1.10. For (a): follow the steps in the proof of Exer. 1.1.9. For (b): start by solving the first two congruences to get an answer modulo 56. Then find a simultaneous solution to that congruence modulo 56 and the third congruence. For (c), the Chinese Remainder Theorem no longer guarantees a solution, since the moduli are no longer pairwise relatively prime. Still, the same procedure will produce a solution if there is one. Be sure to cancel as much as possible.

1.2.1. Factor $z^2 = (x+yi)(x-yi)$ and apply unique factorization in $\mathbb{Z}[i]$. How are the factorizations of $x + yi$ and $x - yi$ into irreducible elements related?

1.4.4. Draw a picture analogous to the right-hand square in Fig. 1.2

1.5.1. As in Figs. 1.2 and 1.3, draw a picture of the fundamental parallelogram with circles of radius 1 centered at the vertices.

1.6.7. Rephrase the problem in terms of the norm in $\mathbb{Q}[\sqrt{319}]$.

1.6.8. Show that such a c fixes a dense subset of $\mathbb{R}$. The continuity of c then implies $c = \mathrm{id}_\mathbb{R}$.

2.2.7. Expand $(a + b)^p$ using the binomial formula, then show that $p \mid \binom{p}{i}$ for $1 \le i \le p - 1$.

2.3.3. Use the First Isomorphism Theorem, Thm. 2.2.6.

2.3.9. For (b): there are at most as many elements in $\mathcal{D}/\mathcal{D}a$ as there are in the set of remainders of division by a. What does Exer. 2.3.8 (b) tell you about the latter set?

2.5.1. Prove the contrapositive: if $I \not\subseteq P, J \not\subseteq P$, then $IJ \not\subseteq P$.

2.5.8. Every integer prime is contained in a prime ideal of $\mathcal{O}$. Show that different primes must lie in different prime ideals.

3.2.4. Prove an intermediate step: $\gamma\mathbb{Z}^2 = \mathbb{Z}^2 \Rightarrow \gamma\mathbb{Q}^2 = \mathbb{Q}^2 \Rightarrow \gamma\mathbb{R}^2 = \mathbb{R}^2$.

M. Trifković, *Algebraic Theory of Quadratic Numbers*, Universitext,
DOI 10.1007/978-1-4614-7717-4, © Springer Science+Business Media New York 2013

3.4.4. Find matrices $M, N \in M_{2\times2}(\mathbb{Z})$ such that $\mathbb{Z}/\mathbb{Z}a \times \mathbb{Z}/\mathbb{Z}b \cong \Lambda_0/M\Lambda_0$ and $\mathbb{Z}/\mathbb{Z}ab \cong \Lambda_0/N\Lambda_0$. Use row and column operations to transform M into N.

4.1.1 For (b): This is a straightforward calculation, but it becomes particularly easy when you work with the basis $\{1, \alpha\}$, in the interesting case $\alpha \in F \setminus \mathbb{Q}$.

4.1.4. By Exer. 4.1.1, we can write $\beta = x + y\alpha$ for some $x, y \in \mathbb{Q}$.

4.1.6. For (a), use the three properties of Def. 2.2.3. For (b), apply f to the equality $(\sqrt{D})^2 = D$.

4.2.14. For (b): $\gamma\Lambda \subseteq \Lambda$ means that we can find $j, k, l, m \in \mathbb{Z}$ such that

$$(*) \qquad\qquad \gamma\alpha = j\alpha + k\beta, \quad \gamma\beta = l\alpha + m\beta.$$

For a suitable $\gamma \in \mathcal{O}_\Lambda$, solve this system for β/α to show that β/α satisfies a quadratic equation over $\mathbb{Q}$. We can re-write $(*)$ in matrix form: $\gamma \begin{bmatrix} \alpha \\ \beta \end{bmatrix} = \begin{bmatrix} j & k \\ l & m \end{bmatrix} \begin{bmatrix} \alpha \\ \beta \end{bmatrix}$. In other words, γ is an eigenvalue of a matrix in $M_{2\times2}(\mathbb{Z})$. Use this to deduce that γ is an algebraic integer.

4.4.4. For a (possibly infinite) list $g_1, g_2, g_3 \ldots \in I$, consider the ascending chain of ideals $\langle g_1 \rangle \subseteq \langle g_1, g_2 \rangle \subseteq \langle g_1, g_2, g_3 \rangle \subseteq \cdots$.

4.4.6. We need an ascending chain of ideals. Try $\ker\varphi \subseteq \ker(\varphi \circ \varphi) \subseteq \ker(\varphi \circ \varphi \circ \varphi)\ldots$.

4.6.6. You can do this without multiplying ideals. First check that $I_i \mid d_1 d_2(\mathbb{Z}(a_1 a_2) + \mathbb{Z}(-b + \delta))$, then compare norms.

4.6.7. Observe that $IK \subseteq J$ by definition. For the reverse inclusion, it's enough to show that $\bar{I}J \subseteq (\mathcal{O} \cdot NI)K$.

4.6.10. For (b) $\Rightarrow$ (c), let $e = [\mathcal{O}_I : \mathcal{O}_c]$, and consider the ideals $J = e\mathcal{O}_c$ and $K = e\mathcal{O}_I$ of $\mathcal{O}_c$. For (c) $\Rightarrow$ (a), use Exer. 4.5.6

4.9.5. Answer: $I = P_1 \cdots P_k(\mathcal{O}n)$, where the prime ideals P_i are all ramified, and $n \in \mathbb{Z}$.

4.9.9. For all $f = 1, \ldots, k$, choose $D_{r_f} \in \mathbb{Z}$ which is divisible by r_f. For all $g = 1, \ldots, l$, choose D_{s_g} which is a non-zero square modulo s_g. For all $h = 1, \ldots, m$, choose D_{i_h} which is a non-square modulo i_h. The Chinese Remainder Theorem produces a $D \in \mathbb{Z}$ satisfying the congruences $D \equiv D_{r_f}$ (mod r_f), $D \equiv D_{s_g}$ (mod s_g), $D \equiv D_{i_h}$ (mod i_h). The field we're looking for is roughly $\mathbb{Q}[\sqrt{D}]$, with extra care needed if one of the primes on the list is 2.

4.9.10. Since $\dim_K L = 2$, the three elements $1, \alpha, \alpha^2$ are linearly dependent.

5.2.1. Use the basis you constructed in Exer. 3.1.1 to compute $A(\Lambda)$. Let S be the circle centered at the origin, with minimal radius r for which Minkowski's Theorem produces a point $(x, y) \in (\Lambda\backslash0)\cap S$. Combine the bound $x^2 + y^2 \leq r^2$ with a congruence on $x^2 + y^2$ (mod p) from the definition of Λ.

5.4.5. You don't need to actually find the solution. Use the fact that 457 is prime, and combine it with Exers. 5.4.1 and 5.4.4.

5.4.6. For (c): Factor $a + \sqrt{399}$ for $a \in \{0, 16, 17, 18, 19, 22\}$. To prove that ideals are nonprincipal, you need to show that equations of the form $x^2 - 399y^2 = n$ have no solution. For that, it suffices to show that n is a non-square modulo one of the factors of $399 = 3\cdot7\cdot19$. For (e): See Exer. 2.3.4.

5.4.8. Look for an example where the class of an ideal dividing 2 has order n. Does this make 2 split, inert, or ramified? Your answer should give you a congruence for D mod 8. Now choose a D so that it's easy to find an element of norm 2^n, but impossible to find one of norm 2^a for $a < n$.

5.4.9. For (a): combine Exer. 5.1.2, Exer. 4.1.4, and Exer. 4.9.5. For (d): let d be the product of all the odd primes among $p_{i_1}, \ldots, p_{i_k}$, and put $D = dd'$. Then the solvability of $N\alpha = p_{i_1} \cdots p_{i_k}$ is equivalent to that of $x^2 - dd'y^2 = 2^e d$, where $e = 0$ if $2 \mid D_F$ but 2 is not among $p_{i_1}, \ldots, p_{i_k}$, $e = 1$ if $2 \mid D_F$ and some $p_{i_j} = 2$, and $e = 2$ if $2 \nmid D_F$ (to justify the last case, see Exer. 4.2.7). Show that, in all cases, we must have $d' = 1$.

6.3.3. It's convenient to treat the cases i even and i odd separately. Since irrationals are trickier than fractions, show that for the purposes of ordering you may replace η by its approximation p_{i+2}/q_{i+2}.

6.6.4. For 469: it turns out that the continued fraction for $\sqrt{469}$ is longer than the one for $(1 + \sqrt{469})/2$, giving more elements of small norm that produce relations. For 577: factor $x + (1 + \sqrt{577})/2$ for $x \in \{-10, -9, -8\}$.

6.7.5. For (d): assume that $N\varepsilon_F = 1$. By Exer. 4.1.4, there exists a $\beta \in \mathcal{O}$ with $\varepsilon_F = \bar{\beta}/\beta$. What does that tell you about the ideal $\mathcal{O}\beta$? What do you deduce when you apply Exer. 4.9.5 to it? How many ramified prime ideals are there in $\mathcal{O}$? Which of them are principal?

7.1.2. Put $I = \alpha P_2$. Look for $\alpha = \beta/\gamma$, where $\beta, \gamma \in \mathbb{Z}[\delta]$ satisfy $N\beta = 54, N\gamma = 4$.

7.1.6. Use Exer. 7.1.5 to find the first few values of each of the forms.

7.2.4. Show that property (7.2.10) implies $q(0) = 0$, then use induction.

7.2.6 Assume that $2 \mid k, j - m, l$. Then $D_F = 4D$ with D square-free and $D \equiv 2, 3 \pmod 4$. We may take $\delta = \sqrt{D}$, so $j = -m$ and $-j^2 - kl = -D$.

7.3.4. The case Ir $\eta_q < 0$ can only happen if $D_F > 0$. In that case, $N\sqrt{D_F} < 0$.

7.4.4. Observe that $GL_2(\mathbb{Z}) = SL_2(\mathbb{Z}) \sqcup \begin{bmatrix} -1 & 0 \\ 0 & 1 \end{bmatrix} SL_2(\mathbb{Z})$.

7.7.4. For (a), generalize Exer. 5.4.9. The key point there is that $\bar{I} = \alpha I$ implies $N\alpha = 1$ (and not -1!). Show that the latter is impossible under the conditions of (c) by generalizing Exer. 6.7.5 (b).

7.10.1. If η is purely periodic, so is $-1/\bar{\eta}$.

Further Reading

1. Bhargava, M.: Higher composition laws. I. A new view on Gauss composition, and quadratic generalizations. Ann. Math. (2) **159**(1), 217–250 (2004)
2. Cassels, J.W.S., Frohlich, A. (eds.): Algebraic Number Theory: Proceedings of an Instructional Conference. Thompson Book, Washington, DC (1967)
3. Cohn, H.: Advanced Number Theory. Dover, New York (1962)
4. Conway, J.H.: The Sensual (Quadratic) Form. Mathematical Association of America, Washington (1997)
5. Cox, D.: Primes of the form $x^2 + ny^2$: fermat, class field theory, and complex multiplication. Pure and Applied Mathematics: A Wiley Series of Texts, Monographs and Tracts Monographs and Textbooks in Pure and Applied Mathematics, vol. 34. Wiley, New York (1997)
6. Harper, M.: $\mathbb{Z}[\sqrt{14}]$ is Euclidean. Canad. J. Math. **56**(1), 55–70 (2004)
7. A. Ya. Khinchin.: Continued Fractions. Dover Books on Mathematics Series. Courier Dover Publications, New York (1964)
8. Lemmermeyer, F.: Binary quadratic forms: An elementary approach to the arithmetic of elliptic and hyperelliptic curves. Available at http://www.rzuser.uni-heidelberg.de/~hb3/publ/bf.pdf
9. Marcus, D.A.: Number Fields. Universitext. Springer, New York (1977)
10. Mollin, R.A.: Quadratics. In: Discrete Mathematics and Its Applications Series, vol. 2. CRC Press, Boca Raton, FL (1996)
11. Serre, J.-P.: A course in arithmetic. In: Graduate Texts in Mathematics, Springer, New York (1973)
12. Zagier, D.B.: Zetafunktionen und quadratische Körper: Eine Einführung in die höhere Zahlentheorie. Hochschultext. Springer, Berlin (1981)

Index

algorithm
  column reduction, 50
  continued fraction
    procedure, 108
  definite reduction of
    forms, 132
  division, 2
  Euclid's, 2
  reduction in $\mathbb{H}$, 160
  row and column
    reduction, 56
associated forms, 176

basis, 45
  oriented, 149
Bhargava cube, 175

cancellation, 34
CF sequence, 108
chain condition
  ascending, 69
Chinese Remainder
  Theorem
  for ideals, 38
  in $\mathbb{Z}$, 7
  via matrices, 58
class number, 87
  narrow, 152
commutative diagram, 155
composition of quadratic
  forms, 172
conductor, 185
conjugate, 23, 61
conjugation (group action),
  147
content, 136
continued fraction

convergent of, 110
elements of, 110, 113
finite, 110
infinite, 113
of a real number, 113
periodic, 116
procedure, 108
purely periodic, 116
convergent, 110
  is best approximation,
    124
  recognizing, 125
  recursion for, 111
cube
  discriminant of, 176

discriminant
  of a quadratic number,
    140
  of a cube, 176
  of a quadratic field, 64
  of a quadratic form, 136
divisibility
  of elements, 33
  of ideals, 36
division algorithm
  in $\mathbb{Z}$, 2
  in $\mathbb{Z}[(1 + \sqrt{-19})/2]$, 36
  in $\mathbb{Z}[i]$, 10
  in $\mathbb{Z}[\omega]$, 19

Eisenstein integers, 17
equivalent forms, 132
equivariant function, 146
Euclid
  algorithm, 2
  domain, 34

lemma, 3, 12
size, 34
strong size, 35

F_+, 150
factorization, 11
  equivalence of, 11
field, 28
First Isomorphism
  Theorem, 30
form, 131
form class group, 172
fractional ideal, 77
  norm of, 78
  of an order, 187
  oriented, 149
Frobenius endomorphism,
  32
fundamental
  domain, 145, 159
  parallelogram, 9, 46
  parallelotope, 94
  unit, 24, 128

Gauss integers, 9
generators, 33
  of $SL_2(\mathbb{Z})$, 164
greatest common divisor
  (g.c.d.)
  in $\mathbb{Z}$, 2
  in $\mathbb{Z}[i]$, 11
group action, 144
  commuting, 146
  equivalence under, 145
  of $GL_2(\mathbb{Z})$ on quadratic
    numbers, 142
  of $GL_2(\mathbb{Z})$ on forms, 142